EAI/Springer Inno and Computing

Series editor
Imrich Chlamtac, European Alliance for Innovation, Ghent, Belgium

Editor's Note

The impact of information technologies is creating a new world yet not fully understood. The extent and speed of economic, life style and social changes already perceived in everyday life is hard to estimate without understanding the technological driving forces behind it. This series presents contributed volumes featuring the latest research and development in the various information engineering technologies that play a key role in this process. The range of topics, focusing primarily on communications and computing engineering include, but are not limited to, wireless networks; mobile communication; design and learning; gaming; interaction; e-health and pervasive healthcare; energy management; smart grids; internet of things; cognitive radio networks; computation; cloud computing; ubiquitous connectivity, and in mode general smart living, smart cities, Internet of Things and more. The series publishes a combination of expanded papers selected from hosted and sponsored European Alliance for Innovation (EAI) conferences that present cutting edge, global research as well as provide new perspectives on traditional related engineering fields. This content, complemented with open calls for contribution of book titles and individual chapters, together maintain Springer's and EAI's high standards of academic excellence. The audience for the books consists of researchers, industry professionals, advanced level students as well as practitioners in related fields of activity include information and communication specialists, security experts, economists, urban planners, doctors, and in general representatives in all those walks of life affected ad contributing to the information revolution.

Indexing: This series is indexed in Scopus, Ei Compendex, and zbMATH.

About EAI

EAI is a grassroots member organization initiated through cooperation between businesses, public, private and government organizations to address the global challenges of Europe's future competitiveness and link the European Research community with its counterparts around the globe. EAI reaches out to hundreds of thousands of individual subscribers on all continents and collaborates with an institutional member base including Fortune 500 companies, government organizations, and educational institutions, provide a free research and innovation platform. Through its open free membership model EAI promotes a new research and innovation culture based on collaboration, connectivity and recognition of excellence by community.

More information about this series at https://link.springer.com/bookseries/15427

Mariya Ouaissa • Zakaria Boulouard
Mariyam Ouaissa • Bassma Guermah
Editors

Computational Intelligence in Recent Communication Networks

Springer

EAI
RESEARCH MEETS INNOVATION

Editors
Mariya Ouaissa
Moulay Ismail University
Meknes, Morocco

Mariyam Ouaissa
Moulay Ismail University
Meknes, Morocco

Zakaria Boulouard
Hassan II University
Mohammedia, Morocco

Bassma Guermah
Parc Technopolis Rabat-Shore
International University of Rabat
Sala Al Jadida, Morocco

ISSN 2522-8595 ISSN 2522-8609 (electronic)
EAI/Springer Innovations in Communication and Computing
ISBN 978-3-030-77187-4 ISBN 978-3-030-77185-0 (eBook)
https://doi.org/10.1007/978-3-030-77185-0

This Springer imprint is published by the registered company Springer Nature Switzerland AG
The registered company address is: Gewerbestrasse 11, 6330 Cham, Switzerland

Preface

Over the past few years, the demand for data traffic has experienced explosive growth thanks to the increasing need to stay online. New applications of communications, such as wearable devices, autonomous systems, drones, and the Internet of Things (IoT), continue to emerge and generate even more data traffic with vastly different performance requirements. With the COVID-19 pandemic, the need to stay online has become even more crucial, as most of the fields, would they be industrial, educational, economic, or service-oriented, had to go online as best as they can. As the data traffic is expected to continuously strain the capacity of future communication networks, these networks need to evolve consistently in order to keep up with the growth of data traffic. Thus, more intelligent processing, operation, and optimization will be needed for tomorrow's communication networks.

The present volume, entitled *Computational Intelligence in Recent Communication Networks*, will focus on the use of AI/ML-based techniques to solve issues related to communication networks, their layers, as well as their applications. It will be a collection of original contributions regarding state-of-the-art AI/ML-based solutions for signal detection, channel modeling, resource optimization, user/application behavior prediction, software-defined networking, communication network optimization, security, and anomaly detection. This book of 15 chapters is an attempt to present solutions to these issues.

In Chap. 1, authors discuss the integration of blockchain with 5G networks and beyond and present how blockchain application in 5G networks and beyond could facilitate enabling various services at the edge and the core. Blockchain was proposed by researches to overcome 5G issues because of its capacities to ensure transparency, data reliability, trustworthiness, and immutability in a distributed environment. Indeed, blockchain has gained momentum as a novel technology that gives rise to a plethora of new decentralized technologies.

Chapter 2 introduces a novel blind detection approach based on Deep Learning Networks (DNN) to efficiently prevent InterSymbol Interferences (ISI) in Offset Quadrature Amplitude Modulation-based Filter Bank Multicarriers (FBMC/OQAM). This approach can provide a better performance than the conventional FBMC/OQAM and reduce the challenges of the application of Multiple-Input Multiple-Output (MIMO) with FBMC/OQAM.

5G New Radio (NR) will use many key technologies to achieve new levels of performance and efficiency. Some of the potential innovations in the 5G physical layer include Non-Orthogonal Multiple Access (NOMA). The main principle of NOMA approach is applying Superposition Coding (SC) in the transmitter side for assigning each sub-carrier to multiple users and Successive Interference Cancellation (SIC) in the receiver side to cancel the other users' signals (interference from other users sharing the same subcarrier). Deep Learning (DL) can provide interesting solutions to the challenges faced by NOMA. In Chap. 3, authors treat the technique NOMA for 5G and how the use of Deep Learning techniques can improve the performance of network.

In autonomous driving, the precise detection of traffic signs is crucial. Accurate detection can dramatically reduce problems such as road accidents involving autonomous vehicles. Numerous algorithms and libraries have been developed to implement reliable and safe detection of traffic signs. In Chap. 4, authors aim to examine the performance of two advanced neural network models, namely Convolutional Neural Networks (CNN) and Recurrent Neural Networks (RNN), and choose the most suitable one in terms of resistance and effectiveness for this particular application.

Chapter 5 aims to analyze the security of vehicular Ad Hoc Networks (VANET), especially the availability. For this analysis, the authors consider the existing trade-off between the rational attacker and the defender. The main goal is to maximize the benefits of attacks and defense strategies. To attain this goal, authors construct an attack-defense tree to model the possible attack and countermeasure that can be assumed. Then they present a detailed analysis based on game theory to search out a Nash equilibrium that reflects the mutual interaction between the attacker and defender.

In Chap. 6, the authors propose a secure and lightweight authentication scheme for Vehicle to Everything (V2X) communication. The proposed solution uses Attribute-Based Signature (ABS) to achieve message authentication, privacy, and integrity, and to enable a secure communication among vehicles on one hand, and between vehicles and Road Side Units on the other hand.

Chapter 7 investigates a novel problem of using unsupervised learning in the task of network intrusion detection in Software Defined Networks (SDN). The authors develop a novel outlier detection method with Isolation Forest (IDS-IF) to effectively detect network anomalies in SDN. The proposed solution not only enhances the detection performance but also reduces the false positive rates as well as computational complexity.

Autonomous Traffic Management (ATM) is a growing field of Intelligent Transportation Systems (ITSs) that aims to replace conventional traffic signals with more efficient cooperative systems. Accordingly, it relays on vehicular communication for sharing awareness messages and hence to enable Autonomous Vehicles (AVs) to make effective control decision to optimize their traffic delay. Nonetheless, little focus has been placed on the insertion of bicycles in the ATM. In Chap. 8, authors propose a Markov Decision Process (MDP) modeling for this scenario and investigate the use of Deep Q-learning to optimize vehicle time loss.

In Chap. 9, the authors address one of the issues related to autonomous vehicles, which is the detection of pedestrians on refuge islands. Their AI-based solution can update the number of pedestrians over time and predict the desired decision.

In Chap. 10, authors use a collection of Arabic YouTube comments that are annotated as either "Hateful" or "Inoffensive" to compare the ability of five Machine Learning algorithms to perform correct classification on hateful Arabic comments. The algorithms are Logistic Regression, Naïve Bayes, Random Forests, Support Vector Machines, and Long-Short Term Memory. The performance metrics are Accuracy, F1-Score, Precision, and Recall.

Chapter 11 introduces an idea based on the presentation of service composition approach guided by the user-intention and the context-awareness, in addition to a Context-Intentional Service Composition Architecture (CISCA) for implementing this approach. Authors will present an Artificial Intelligence (AI) planning technique to solve a composition problem, namely the Planning Domain Definition Language (PDDL) and the basis for reciprocal transformation with the Web Ontology Language (OWL).

Chapter 12 introduces a combination between Genetic Algorithm (GA) and several Machine Learning Algorithms (MLA). This combination can offer a better feature selection of molecular descriptors in a Quantitative Structure-Activity Relationship (QSAR) classification and prediction problem. The authors tested their approach on a dataset of anti-Human Immunodeficiency Virus (anti-HIV) molecules.

Biomedical datasets have a particular focus, such as diagnosing, preventing, or dealing with diseases while facing pressure to reduce costs and improve efficiency. Nevertheless, many unspecific signs and symptoms can make the doctors' tasks more difficult and sometimes result in undesirable errors to rank the adequate disease from the patient records and thus affect the quality of services. Chapter 13 develops a ranking disease system with enhanced accuracy to solve this problem.

In Chap. 14, authors develop a ranking-based clustering strategy alongside two deep neural networks. The first neural network will focus on the market state representation, while the second will model the period by which the ranking mechanism is applied. The analysis suggests that COVID-19 is having an unprecedented influence on markets. The investment process needs to be continuously adjusted using different indices for more synthesized patterns. This paper offers a significant analysis of investors' evaluation of the overall economy's consequences, specifically the stock markets.

Chapter 15 offers a solution to the overcrowded hospital emergency department. This solution, based on the Artificial Immune Systems algorithm, provides an effective optimization of the incoming patients' flux.

Contents

1 **An Overview of Blockchain and 5G Networks** 1
Hajar Moudoud, Soumaya Cherkaoui, and Lyes Khoukhi

2 **Deep Learning Approach for Interference Mitigation in MIMO-FBMC/OQAM Systems** 21
Abla Bedoui and Mohamed Et-tolba

3 **Deep Learning-Based Non-orthogonal Multiple Access for 5G and Beyond Networks** 39
Mariyam Ouaissa, Mariya Ouaissa, Zakaria Boulouard, and Sarah El Himer

4 **Traffic Sign Detection: A Comparative Study Between CNN and RNN** 53
Amal Bouti, Mohammed Adnane Mahraz, Jamal Riffi, and Hamid Tairi

5 **Merging Attack-Defense Tree and Game Theory to Analyze Vehicular Ad Hoc Network Security** 69
Meriem Houmer, Mariya Ouaissa, Mariyam Ouaissa, and Moulay Lahcen Hasnaoui

6 **A Secure Vehicle to Everything (V2X) Communication Model for Intelligent Transportation System** 83
Mariya Ouaissa, Mariyam Ouaissa, Meriem Houmer, Sara El Hamdani, and Zakaria Boulouard

7 **A Novel Unsupervised Learning Method for Intrusion Detection in Software-Defined Networks** 103
Zakaria Abou El Houda, Abdelhakim Senhaji Hafid, and Lyes Khoukhi

8 **Deep Reinforcement Learning Modeling of a V2V Communication-Based Bike Avoidance Protocol for Increased Vehicular Flow** ... 119
Sara El Hamdani, Salahedine Loudari, Mariyam Ouaissa, Mariya Ouaissa, and Nabil Benamar

9 **Deep Learning-Based Modeling of Pedestrian Perception and Decision-Making in Refuge Island for Autonomous Driving** ... 135
Badr Ben Elallid, Sara El Hamdani, Nabil Benamar, and Nabil Mrani

10 **Machine Learning for Hate Speech Detection in Arabic Social Media** ... 147
Zakaria Boulouard, Mariya Ouaissa, and Mariyam Ouaissa

11 **PDDL Planning and Ontologies, a Tool for Automatic Composition of Intentional-Contextual Web Services** ... 163
Abdelmajid Daosabah, Hatim Guermah, and Mahmoud Nassar

12 **QSAR Anti-HIV Feature Selection and Prediction for Drug Discovery Using Genetic Algorithm and Machine Learning Algorithms** ... 191
Houda Labjar, Najoua Labjar, and Mohamed Kissi

13 **Mining Electronic Health Records of Patients Using Linked Data for Ranking Diseases** ... 205
Siham Eddamiri, Elmoukhtar Zemmouri, and Asmaa Benghabrit

14 **The COVID-19 Pandemic's Impact on Stock Markets and Economy: Deep Neural Networks Driving the Alpha Factors Ranking** ... 219
Badr Hirchoua, Brahim Ouhbi, and Bouchra Frikh

15 **An Artificial Immune System for the Management of the Emergency Divisions** ... 245
Mouna Berquedich, Ahmed Chebak, Oualid Kamach, Oussama Laayati, and Malek Masmoudi

Index ... 265

About the Editors

Mariya Ouaissa is a research associate and practitioner with industry and academic experience. She holds a Ph.D. and has graduated in 2019 in computer science and networks from the Laboratory of Modelisation of Mathematics and Computer Science at ENSAM-Moulay Ismail University, Meknes, Morocco. She is a networks and telecoms engineer, graduated in 2013 from the National School of Applied Sciences Khouribga, Morocco. Dr. Ouaissa is co-founder and IT consultant at IT Support and Consulting Center. She was working for the High School of Technology of Meknes Morocco as a visiting professor from 2013 to 2020. She is member of the International Association of Engineers and International Association of Online Engineering, and since 2021, she is an "ACM Professional Member." Dr. Ouaissa has made contributions in the fields of information security and privacy, Internet of Things security, and wireless and constrained networks security. Her main research topics are IoT, M2M, D2D, WSN, cellular networks, and vehicular networks. She has published over 20 papers (book chapters, international journals, and conferences/workshops) and a book entitled *Proceedings of International Conference on Advances in Communication Technology, Computing and Engineering*. Dr. Ouaissa has served and continues to serve on executive and technical program committees and as a reviewer of numerous international conference and journals such as *IEEE Access* and *Wireless Communications and Mobile Computing*. She is the general chair of the ICACTCE'21 Conference.

Zakaria Boulouard is currently a professor in the Department of Computer Sciences, Faculty of Sciences and Techniques of Mohammedia, Hassan II University, Casablanca, Morocco. In 2018, he joined the Advanced Smart Systems research team at the Computer Sciences Laboratory of Mohammedia. He received his Ph.D. degree in 2018 from Ibn Zohr University, Morocco, and his engineering degree in 2013 from the National School of Applied Sciences, Khouribga, Morocco. His research interests include artificial intelligence, big data visualization and analytics, and optimization and competitive intelligence. Since 2017, he is a member of "Draa-Tafilalet Foundation of Experts and Researchers," and since 2020, he is an "ACM Professional Member." He is general co-chair of the ICACTCE'21 Conference.

He is a co-editor of a book entitled *Proceedings of International Conference on Advances in Communication Technology, Computing and Engineering*.

Mariyam Ouaissa is a Ph.D. research associate and consultant trainer in computer science and networks at Moulay Ismail University, Meknes, Morocco. She received her Ph.D. degree in 2019 from the National Graduate School of Arts and Crafts, Meknes, Morocco, and her engineering degree in 2013 from the National School of Applied Sciences, Khouribga, Morocco. She is a communication and networking researcher and practitioner with industry and academic experience. Dr. Ouaissa's research is multidisciplinary that focuses on the Internet of Things, M2M, WSN, vehicular communications and cellular networks, security networks, congestion overload problem and the resource allocation management, and access control. She is serving as a reviewer for international journals and conferences including as *IEEE Access* and *Wireless Communications and Mobile Computing*. Since 2020, she is a member of the International Association of Engineers IAENG and International Association of Online Engineering, and since 2021, she is an "ACM Professional Member." Dr. Ouaissa is the TPC chair of the ICACTCE'21 Conference. She is a co-editor of a book entitled *Proceedings of International Conference on Advances in Communication Technology, Computing and Engineering*.

Bassma Guermah is a professor in the Computer Science Engineering School at the International University of Rabat (UIR) and a member of TICLab. She obtained her doctorate in computer science and telecommunications from the National Institute of Posts and Telecommunications (INPT) in 2018. She obtained her degree in software engineering from the National Institute of Statistics and Applied Economics (INSEA) in 2014. Her research activities revolve around machine learning/deep learning (artificial intelligence), signal processing, robotics, context-aware service-oriented computing, ontologies, and semantic web.

Chapter 1
An Overview of Blockchain and 5G Networks

Hajar Moudoud, Soumaya Cherkaoui, and Lyes Khoukhi

1.1 Introduction

Fifth-generation (5G) networks and beyond are expected to enable a wide range of applications from the industrial internet of things, virtual reality, autonomous driving to real-time gaming [1] will enable them to benefit from massive machine-like communications, improve mobile broadband, and provide ultra-reliable, low-latency communications. The 5G road map aims to provide users with up to 10 Gbps, 1000 more network capacity, 10 Tbps per square kilometer, and 1-ms latency [2]. To achieve these requirements, several technologies have been proposed among them: cloud computing (CC), multi-access edge computing (MEC), software-defined network (SDN), and network function virtualization (NFV) [3]. Besides the new technologies, 5G will incorporate new radio access techniques including massive MIMO and millimeter wave, and D2D connectivity to 5G cellular networks.

The 5G networks will support new technologies and offer new services. These technologies, however, will cause several challenges in terms of users' security, data privacy, and integrity. Unlike centralized cellular networks (i.e., 3G, 4G, etc.), 5G networks are expected to distribute and decentralize services that emphasize security

H. Moudoud (✉)
Université de Sherbrooke, Sherbrooke, QC, Canada

University of Technology of Troyes, Troyes, France
e-mail: hajar.moudoud@usherbrooke.ca

S. Cherkaoui
Université de Sherbrooke, Sherbrooke, QC, Canada
e-mail: soumaya.cherkaoui@usherbrooke.ca

L. Khoukhi
Normandy University, Rouen, France
e-mail: lyes.khoukhi@ensicaen.fr

M. Ouaissa et al. (eds.), *Computational Intelligence in Recent Communication Networks*, EAI/Springer Innovations in Communication and Computing,
https://doi.org/10.1007/978-3-030-77185-0_1

challenges [4]. Specifically, the security management in 5G networks is complex because it operates in a flexible and dynamic environment where a massive number of devices are connected.

For secure, transparent, and immutable 5G networks, there is a need to ensure security while exploiting new technologies such as distributed ledger technologies (DLT), artificial intelligence (AI), and machine learning (ML). Among the existing technologies, blockchain is a promising technology that has the potential to revolutionize the way services are offered 5G networks [5]. Blockchain technology is a distributed, immutable, and secure ledger that ensures trust among unreliable entities without relying on a centralized third party. Blockchain has the capacity to be merged with the 5G networks to provide reliable resource sharing, secure storage, smart authentication, and security management [6]. Consequently, blockchain with its inherent features will provide massive communication in a distributed environment while ensuring high security, data privacy, and reliability.

Currently, one of the challenging aspects of 5G networks is the need to assure transparency, immutability, and decentralization for its large number of users and services. Blockchain technology, with its inherent properties will enable secure and trustworthy data transactions in a P2P manner for various services/users in the 5G network [7, 8]. Therefore, blockchain integration with 5G networks will result in self-managing, self-securing, and self-maintaining networks without the need for a central authority. In fact, 5G is expected to provide a connection for a large number of devices with resources and services (network slicing). In this regard, distributed blockchain with its security features will enable a new generation of massive communication by providing transparent provisioning between devices.

In addition to the benefits of integrating blockchain with 5G networks, 5G will enable a wide range of blockchain-based applications such as autonomy's resource, automated management, and reliable content base storage [9, 10]. For instance, blockchain can be leveraged by cellular service providers (CSPs) to enable future services for mobile industries. In order to automatically handle transactions in CSPs, blockchain features such as smart contracts which are self-executing code will enable a secure and automated sharing of infrastructures. Furthermore, the blockchain and ML empowered intelligent 5G beyond network will enable a new stack of technologies that will empower self-aggregating communication and intelligent resource management beneficial to 5G networks.

1.2 Background

In this section, first, we provide a brief background on the blockchain and its features. Next, we provide a brief description of blockchain features.

1.2.1 Blockchain

Blockchain is a transparent technology that establishes trust among unreliable entities [10, 11]. This decentralized technology ensures secure information transmission without the interference of a third authority. Blockchain is a distributed ledger that keeps a continuously growing set of data records called blocks. Each block contains a collection of transactions committed by members of the blockchain. A blockchain transaction is a piece of information meant to be stored in a safe database [12, 13]. For instance, a transaction could contain technical metadata of a mobile (i.e., size, type, or timestamp). The blockchain is an immutable ledger that permits transaction verification by members who could be dishonest. To add a new block to the blockchain, all members of the chain should reach an agreement; this is referred to as a consensus. Once a consensus is established among all members, the new block is validated and then added to the chain.

Blockchain technology has many characteristics, such as being decentralized, immutable, and having no single point of failure (SPOF). Blockchain can operate in a decentralized environment, and each member of the chain has an integral copy of the ledger, meaning data is stored in a peer-to-peer environment [14]. This redundancy of information guarantees data non-repudiation, making it difficult to cause major disruption. Blockchain is designed to be immutable; once a block is added to the chain, any changes inside the block will be extremely difficult. The blockchain enables several technologies like the hash function, the digital signature, and the timestamp. If a malicious user wants to change a block, this will cause the hash to change also, meaning he needs to reach a new agreement for this block and other blocks following it. Due to the distributed and shared nature of the blockchain, the ledger cannot be controlled by a centralized entity, meaning it has no SPOF.

1.2.1.1 Blockchain Taxonomy

A blockchain is composed of a family of blocks linked together by a hash. The hash function is used to map data stored in a transaction and generates a unique fingerprint [15]. The hash does not only depend on the new transaction but on the previous transactions also. A new transaction is broadcast to miners of the chain and waits to be confirmed. To verify a transaction, miners need a digital signature to certify the authenticity and integrity of the transaction. The blockchain uses the Elliptic Curve Digital Signature Algorithm to sign a transaction. Once the blockchain miners approve a transaction, it is written into a block. A block is added to the chain when the consensus is established, and the block has reached a certain number of verified transactions. Each block refers to the previous block and together forms the blockchain. Figure 1.1 presents how blocks are linked together.

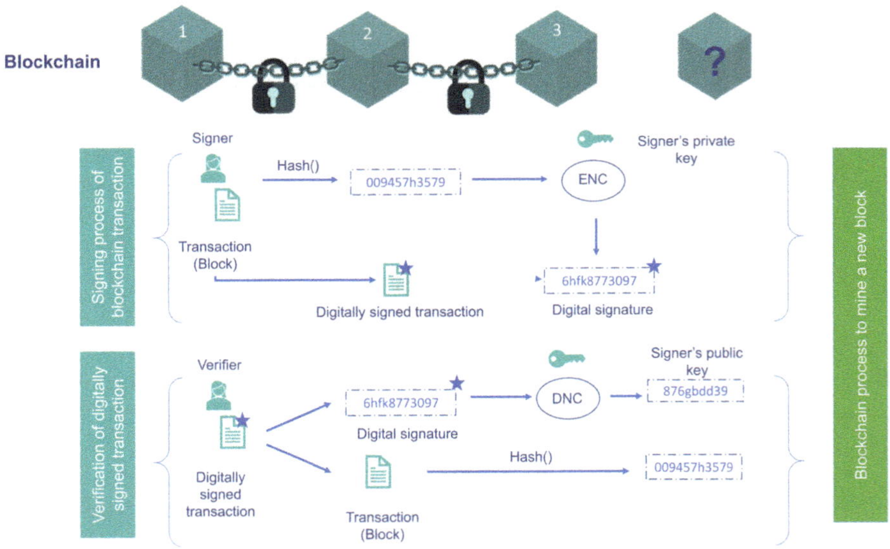

Fig. 1.1 Blockchain workflow

1.2.1.2 Blockchain Platform Types

Current blockchain networks can be classified into three types: public, private, and consortium blockchain [16]. A public blockchain is permissionless where all members can join the network. It is publicly open for members to read, write, or validate a transaction without the approval of third parties. In a private blockchain, only members can participate in the network, meaning it is centralized. The owner of the consortium blockchain, only a group of authorization members, can validate transactions [17]. These members can be chosen in advance.

1.2.1.3 Blockchain Consensus

Blockchain consensus is a protocol that establishes the core of the blockchain and decides how the agreement should be reached among miners to append a new block to the chain [18]. Blockchain consensus can be defined as an algorithm that helps the distributed network to make decisions, while others describe the consensus algorithm as the mechanism that brings about all nodes to take a decision about the same transaction either to add it to the chain or reject it. According to several researchers making a consensus among unreliable miners is a transformation of the Byzantine generals. There are several types of blockchain consensus. The most common consensus algorithms are Proof of Work (PoW), Proof of Stake (PoS), Proof of Activity (PoA), and Practical Byzantine Fault Tolerance (PBFT).

1.2.1.4 Blockchain Smart Contract

Blockchain smart contracts are self-executing programs non-modifiable implemented inside the chain that is intended to automatically execute actions to the terms of a contract [19]. Smart contracts are implemented permanently inside the blockchain to execute a code once some conditions are verified. The most used platform to run smart contracts is the Ethereum blockchain. These smart contracts are coded using a coding language called Solidity. For example, blockchain smart contracts can be implemented to help to manage a system or to establish access control inside a network. The smart contracts can run dynamically in the network.

1.2.1.5 Blockchain Sharding

The sharding consists of partitioning a large collection across several servers, which allows for distributed management of the collection and thus improves scalability [20]. Blockchain sharding refers to the artificially dividing of the workload of the transaction procession into a single shard; in this way, a single transaction can be validated and stored by many members working in parallel. Yet, applying sharding to the blockchain presents several challenges, such as weak security, additional throughput, and increased risk for data loss.

1.2.1.6 Blockchain Oracle

Blockchain cannot access external data from the network. This is where the blockchain oracle intervenes. It is a service provider (trusted third party) that verifies the data authenticity and attests to the facts in order to bring outside world data into the chain.

1.3 5G Networks and Beyond: An Overview

In this section, we present an overview of the 5G networks.

1.3.1 Software-Defined Networking (SDN)

SDN enables external control of data away from network hardware to software referred to as the controller. The controller manages packet flow to provide intelligent networks. With the controller, users will be able to manage network equipment using software, introducing new services. A smart 5G network will enable new operations and provide new services on demand while ensuring efficiency. 5G SDN will be able to control and orchestrate services in a seamless and efficient manner. The SDN architecture will offer great flexibility to 5G networks allowing it to adapt perfectly to the dynamic nature of 5G bandwidth.

1.3.2 Network Function Virtualization (NFV)

NFV refers to the replacement of hardware infrastructure by virtualization software and for different network functions (e.g., VPN, load balancers, firewall, routers, switches). NFV decouples the network functions from physical infrastructure and permits it to run virtually on a cloud infrastructure. The key objective of NFV is to transform the way networks are built and services are delivered. In 5G, NFV will enable the distributed cloud, helping leverage scalability, dynamicity, and flexibility. The SDN and NFV are complementary to each other and both help network abstraction and virtualization. The difference between NFV and SDN is that NFV isolates network control functions from network forwarding functions, whereas NFV isolates the network functions from physical infrastructure to the cloud.

1.3.3 Network Slicing

Network slicing enables multiple networks to work virtually over one physical network infrastructure. Each network slice is isolated from the physical network to meet the requirements requested of an application. Integrating NFV with network slicing will allow multiple applications/services to be deployed for users. Consequently, applications or services running over a network slice can display a high quality of experience (QoE) and high-quality services (QoS) beneficial for users.

1.3.4 Multi-Access Edge Computing (MEC)

MEC reduces network congestion and thus achieves a faster response. It enables computing capabilities of the cloud to the network edge that allows data to be processed near to the devices and thus users. Furthermore, by enabling the computing locally, MEC reduces significantly the energy need to process data and storage space. MEC enabled services, applications, and operations to be closer to the users, which enhances the QoS for users; it enables network collaboration that improves the quality of experience.

1.3.5 Device to Device (D2D)

In current cellular networks, base stations are responsible for establishing communications between two devices. These communications are going through base stations even if the two devices are in the same ranges. Consequently, the spectral efficiency is low because of the delay added, which is unsuitable for real-time applications. To solve this issue, the 5G networks proposed using D2D communications to allow two devices close to each other to communicate using a direct link. This concept creates multi-hop relays among several devices that increase the data rate and improve QoS. Furthermore, D2D connectivity will assist 5G networks to be more malleable in terms of offloading and energy efficient because it eliminates unwanted traffic from the core network.

1.3.6 Cloud Computing (CC)

CC was proposed to achieve the ever-growing requirements of 5G networks, such as resource orchestrations, data storage, and mobile sensing. CC enables resource offloading by virtualizing physical infrastructure to dynamically provide 5G services with their requests. Cloud computing includes two tiers: (1) Infrastructure Providers that manage the physical infrastructures (InPs) and (2) Service Providers (SPs) that provide services to network users.

1.4 Blockchain for 5G

In this section, we discuss the benefits at various levels of blockchain integration with 5G networks. Then, we present a taxonomy of the opportunities brought by blockchain applications on 5G (Fig. 1.2).

1.4.1 Blockchain Integration with 5G Networks

5G and blockchain are two prominent technologies that could reshape the future of technologies and the telecommunication sector. On one hand, blockchain can provide a distributed environment to secure 5G services, operations, and data. Blockchain techniques can assist 5G to enable features such as decentralization, immutability, and transparency. On the other hand, 5G networks are highly distributed and require enabling new technologies such as NFV, SDN, D2D, MEC, and CC. These technologies are complex to orchestrate and manage. Furthermore, 5G networks share resources, services, and operations among several stakeholders that

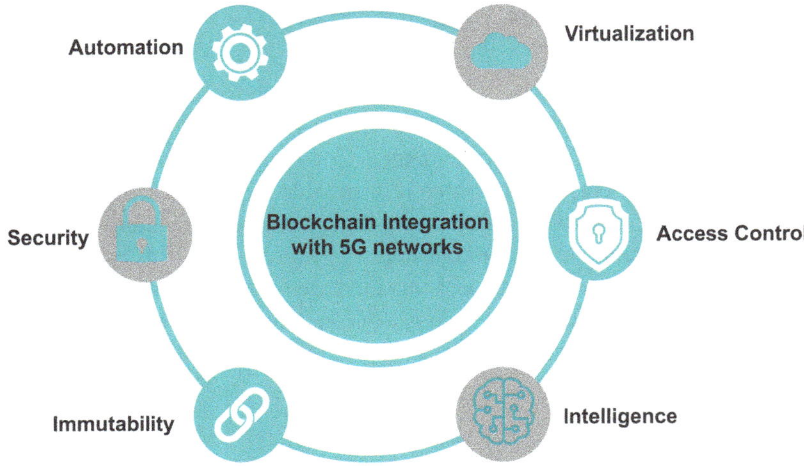

Fig. 1.2 Blockchain integration with 5G opportunities tree

could be dishonest. Consequently, blockchain will enable future 5G networks with a high level of security, coordination, and manageability required among 5G users. All these new features will present challenges that need to be solved for the safe deployment of 5G networks.

The reason behind the integration of blockchain with 5G networks comes for the most part from the prominent features of blockchain that could solve the challenges in 5G networks in terms of security, privacy, management, and transparency. Because 5G involves several stakeholders, it is difficult to leverage trust while preserving their security. With the use of blockchain technology, 5G stakeholders are secured without the need of trusting each other. Blockchain consensus establishes trust among unreliable entities without the use of a trusted third party. Furthermore, shared services and operations are secured using the blockchain immutable ledger. Indeed, once data is stored inside the blockchain ledger, it cannot be altered or falsified, and blockchain uses cryptographic signature and hash function to secure data. Moreover, blockchain smart contracts can be used to automate the management of 5G resources and services. The entire process is transparent, reliable, and decentralized. Figure 1.3 presents a conceptual diagram for the integration of blockchain with 5G networks. Next, we present the opportunities brought by the integration of blockchain with 5G.

1.4.2 Opportunities Brought by Blockchain Integration with 5G Networks

Blockchain interesting features will provide a new set of solutions to improve 5G networks in terms of security, transparency, immutability, privacy, and interoper-

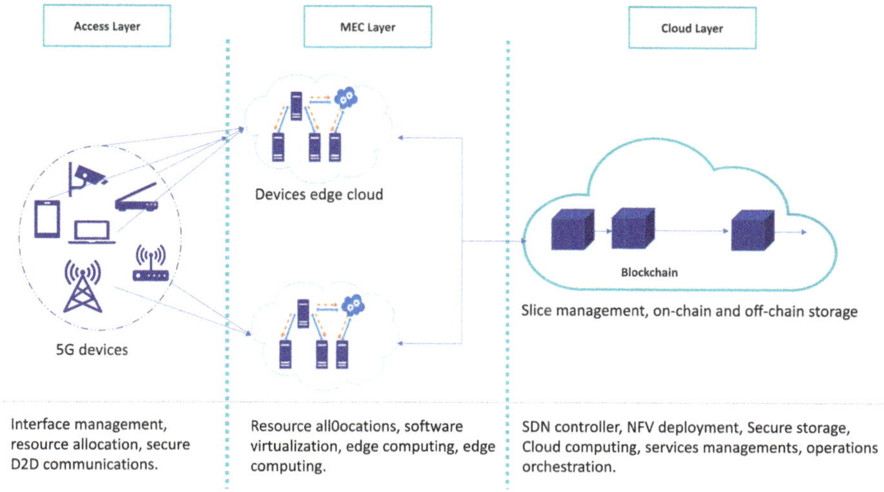

Fig. 1.3 Blockchain merging with 5G networks

ability. Consequently, 5G networks should take advantage of blockchain to improve their flexibility, manageability, and security of network services and operations. Next, we present the opportunities that blockchain technology can conduct to 5G networks with regard to security improvements and performance enhancements.

1.4.2.1 Security Improvements

5G networks face several security challenges, from data privacy to authentication vulnerability such as frequent authentication applied to ultra-dense networks [21]. These challenges expose 5G networks to attacks and increase their impact. Especially, when services and operations are shared among several stakeholders when one service is under attack, all the stakeholders taking part are exposed to attacks. Fortunately, blockchain features will improve 5G networks in terms of security, data privacy, and access control. For example, blockchain Byzantine Fault Tolerance (BFT) consensus can help 5G networks to achieve trust in a distributed network even when some of the stakeholders in the network respond with incorrect information. Furthermore, blockchain can decentralize network management without the need for a third-party authority. For instance, the use of blockchain-based cloud computing can enable the decentralization of MEC 5G networks that take out the control from the core network, provide decentralize management, eliminate SPF issues, and improve trust in the network. In addition, blockchain can help to secure D2D communication by building a centralized peer-to-peer blockchain network, which considers each device as a blockchain miner that holds a copy of the ledger, verifies the authenticity of a transaction, and monitors transactions for better system reliability.

5G networks, with high connectivity and low latency, come to support distributed new services and operations, which need decentralized management, limited access control, secure data sharing, and authentication services. Blockchain with smart contracts can establish a decentralized access control for both stakeholders and services. Because blockchain uses computing power to establish trust, data sharing, resource allocation, and spectrum sharing can be strongly secured against attacks such as False Data Injection or Modification Data Attacks. Indeed, several researchers have proven the efficiency of blockchain technology to secure 5G networks in terms of better access control using blockchain contracts, secure data sharing using hash functions, and cryptographic signatures. Furthermore, blockchain can host several technologies like Artificial Intelligence (AI) that could work on their own. Applied on 5G networks, blockchain and AI can be able to operate in an independent manner to ensure better network orchestration and service management. All without the need of a trusted third authority.

With the support of blockchain features, 5G networks can fully benefit from better access control to services, secure data storage, and improved network management using blockchain smart while ensuring transparency and privacy. Furthermore, the strong immutability of blockchain ledger can provide a high degree of security for 5G users especially when sensitive data are being shared at a high speed among several 5G stakeholders. Furthermore, blockchain smart contracts can provide efficient access control mechanisms and authentication solutions that could help improving authentication issues in 5G networks. Blockchain smart contracts can implement automated access rules that will help authenticate 5G stakeholders without relying on external authority, can limit malicious behaviors in the network, and can detect threats without leaking stakeholders' information. Besides, because blockchain uses hash functions, data is signed making it difficult to corrupt or falsified. Blockchain is capable of proving high data protection when sharing among untrustworthy 5G stakeholders while ensuring transparency.

1.4.2.2 Performance Enhancements

Blockchain technology can improve the performance of 5G systems. First, blockchain miners can verify data access and credentials that help 5G networks protecting data. Second, blockchain smart contracts can manage automatically 5G services with a low latency, which may lead to reduce the management costs. Finally, the use of a decentralized blockchain can provide a flexible, efficient, and secure data delivery system suitable with the 5G complex environment. For instance, blockchain can leverage a peer-to-peer network to secure communication among all 5G stakeholders and ensure trust among participants using appropriate consensus algorithms. Overall, blockchain integration with 5G networks can reduce communication costs and latency and provide a secure platform for data exchange for all stakeholders, all of which improves the overall system performance.

Interference management is a known factor that influences the performance of wireless mobile networks. Because of the dense deployment in 5G networks and cellular interference, interference management will be further complex and hard to overcome. Further, collaborative communications will play an important part in 5G networks, managing a huge number of resources, services, and stakeholders while ensuring that a high quality of services will even be greater. In this context, blockchain can solve some of these issues by decentralized interference administration. For example, blockchain can implement a cooperative interference management algorithm in blockchain smart contracts to ensure mutual trust and coordination protocols of interference.

1.5 A Scalable and Secure Blockchain Suitable for 5G

We solely reviewed the integration of blockchain with 5G networks. In this section, we present our vision of merging blockchain with 5G. We provide an overview of the proposed architecture named Block5G.

1.5.1 A Scalable and Secure Blockchain Architecture Suitable for 5G

Following our presentation of blockchain opportunities that blockchain can bring to 5G networks, in this section, we present our vision of a blockchain integration with 5G networks. 5G network needs a secure, scalable, and non-computing intensive solution that meets the requirement of low-latency, high communication rate provided by the network. In this context, we present a scalable and secure blockchain architecture that uses the sharding concept, and blockchain oracles suitable with 5G requirements named Block5G.

1.5.2 Architecture

As illustrated in Fig. 1.4, we consider data exchange in a 5G network where a large number of unreliable users are connected through a D2D communication. We propose a scalable, secure, and trustworthy blockchain architecture composed of three layers, as follows:(1) the access layer that includes 5G devices that send and receive data; (2) the edge layer that is responsible for forwarding packets and verifying their validity using a blockchain consensus and blockchain oracles; and (3) cloud layer that is responsible for storing the data and scaling the blockchain while keeping security guarantees.

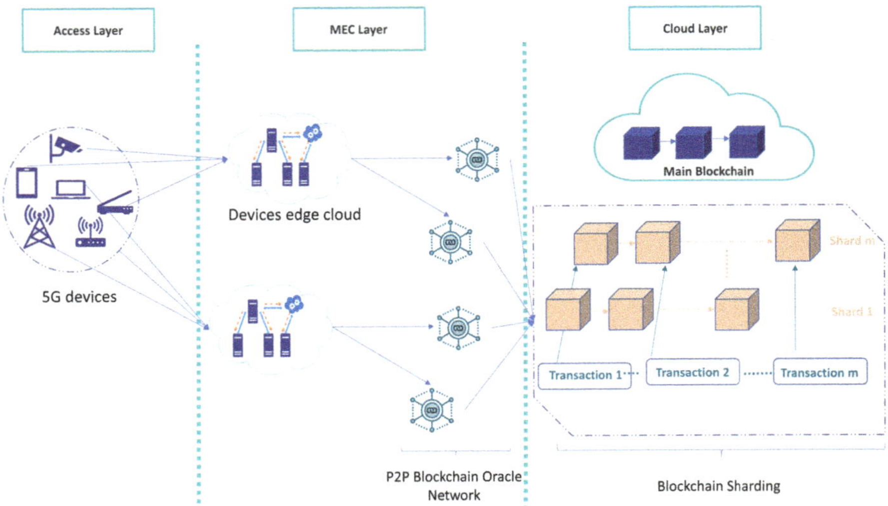

Fig. 1.4 Scalable, trustworthy, and secure blockchain suitable with 5G

1.5.2.1 Shared Blockchain

The scalability has been a core problem in the integration of blockchain within 5G. This is because the 5G includes a large number of devices and users that communicate at a high rate and thus generate big data. To enable horizontal scalability, blockchain sharding was proposed, and it consists of partitioning each transaction into several shards and processing it independently. In this chapter, we propose building clusters with multiple nodes (i.e., miners) to process each shard in parallel. Figure 1.5 shows the three main components of the Block5G sharding concept, as follows: (1) the main chain; (2) the master clusters; and (3) slave devices. The master cluster contains n slave devices that could be either honest or dishonest. Let $T_i(x)$ represent the i_{th} transaction in a block. To enable the sharding, each transaction is divided into n disjoint portions. All master clusters are responsible for verifying a portion of the transaction $t_{i,j}(x)$, where i is the transaction number on a block and j is the cluster number. All slave devices could verify a transaction portion, and output 0 or 1 indicates invalid and valid transactions, respectively. Let n denote the total number of slave devices in a master cluster j that can tolerate up to $t \leqslant \frac{n}{3}$ dishonest devices. We define a transaction verification process as follows:

$$T_i(x) = \frac{\sum_{k=0}^{m} t_{i,j}(x)}{m}, \tag{1.1}$$

where m is the total number of miners that have participated in the verification process.

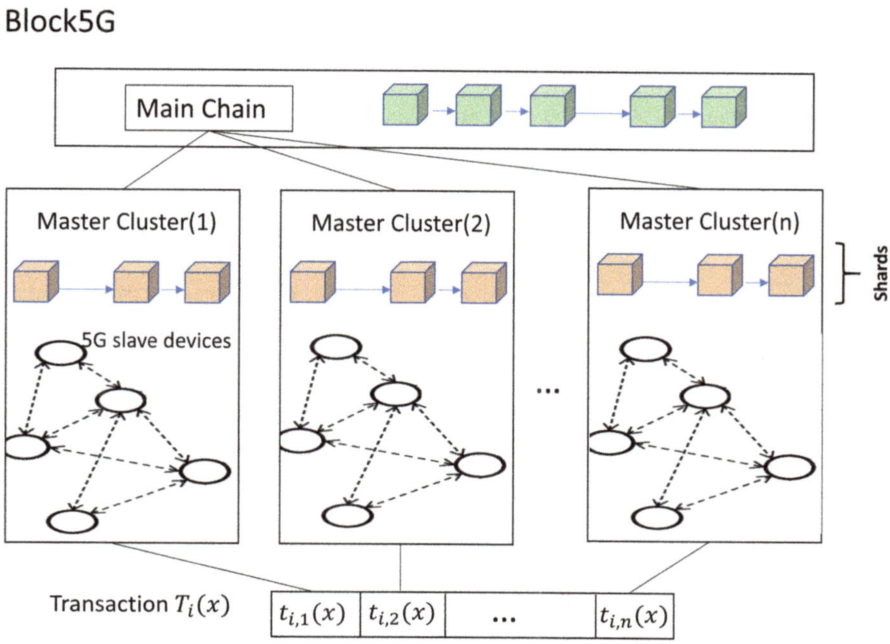

Fig. 1.5 Scalable and trustworthy blockchain sharding in 5G context

The transaction verification process has two possible outcomes: valid and invalid, each of which carries its own reward structure. As compensation for their efforts, the clusters are awarded (R) whenever they verify a transaction correctly. In the case of correct verification, a cluster reward is as follows:

$$R(T_{i,j}) = \frac{R(t_{i,j}(x))}{R(T_{i,Total}(x))}, \tag{1.2}$$

where $R(t_{i,j}(x))$ and $R(T_{i,Total}(x))$ represent, respectively, the reward for a valid transaction portion and total reward for a valid transaction. Note that the cluster trust value is only used to determine rewards and penalties and does not necessarily correspond to the cluster output. Consequently, we enable secure sharding and prevent data loss and incentive clusters to behave honestly.

1.5.2.2 Peer-to-Peer Oracle Network

We propose using a peer-to-peer (P2P) oracle network to verify the data queries and authenticate its source. Blockchain cannot access external data of the network. This is where the blockchain oracle interferes, and it is a service provider (trusted third party) that verifies the data authenticity. However, trusting a single third party may lead to providing corrupt or inaccurate data. To this end, we propose using a P2P oracle network that ensures the truth value of 5G data. We assume each received data sent from a 5G device could be either valid V or false F. There is m oracle in the P2P network and only n oracle verify the data. For each oracle, $o \in$ [n,m] has a q probability that data d is correct about a given proposition.

$$O_i(d) = \begin{cases} q & \text{when data is valid} \\ 1-q & \text{when data is false} \end{cases} \tag{1.3}$$

The value of $o_i(d)$ is independent of $o_j(d)$ for all $i \neq j$. In other words, each oracle's trust values are independent of other oracles in the network trust values. Furthermore, oracles need to place a deposit to participate in a randomly chosen verification process. First, the oracle submits its verification probability $q \in$ [0,1] about a data d. Then, this probability is applied to a trust weight $W \in$ [0,1] that is, informally, the parameter within the network that verifies the input data within the network's hidden layers. Formally:

$$W_i = \frac{\alpha_i}{\beta_i}, \tag{1.4}$$

where α_i is the sum of corrected verification performed and β_i is the total number of verifications performed.

$$V(d) = \sum_{i=0}^{n} o_i(d) \times W_i. \tag{1.5}$$

Once a data verification process $V(d)$ has accumulated sufficient verifiability during a maximum of a period of time $\delta(t)$, it is decided. This period of time is a fixed value decided by network operators. The verification process has three possible outcomes: (1) valid (V), if $V(d)$ value to be strictly positive, (2) false (f), if $V(d)$ value to be strictly negative, and (3) undefined (U) if $V(d)$ value is equal to zero. In this last case, the data verifiability process is only assigned to the oracle with the highest trust weight.

Broadly speaking, oracles are rewarded when they participate in a verification process and their verification probability matches it. Conversely, those who gave incorrect verification are penalized. In the case of undetermined outcomes, oracles receive no rewards or penalties. As argued in this chapter, the proposed data verification process incentivizes the oracles to behave honestly on the validity of data.

1.5.3 Design Components

The transaction verification procedure of our Block5G of the following: (1) Initialization and (2) Reward. This procedure starts with the initialization and then proceeds in periods to the rewarding phase. We now explain each component in more detail.

1.5.3.1 Initialization

The initial set of participants (i.e., nodes) are invited to provide a deposit ($stake$) to validate or endorse a transaction. That is, a node is given the chance to verify a transaction chosen uniformly at random from the unverified transaction pool. The deposit is placed before the verification process. Because the nodes are grouped in clusters, the outcome of the transaction verification reward is a function of the sum of the total transaction verification reward weighted by the cluster value in the cluster and the deposits.

1.5.3.2 Reward

Broadly speaking, nodes (i.e., miners) are rewarded for transaction verification in which they took part. Conversely, those who provided false verification are penalized. The steps of reward/penalize are depicted in Fig. 1.6. Let r_i denote the

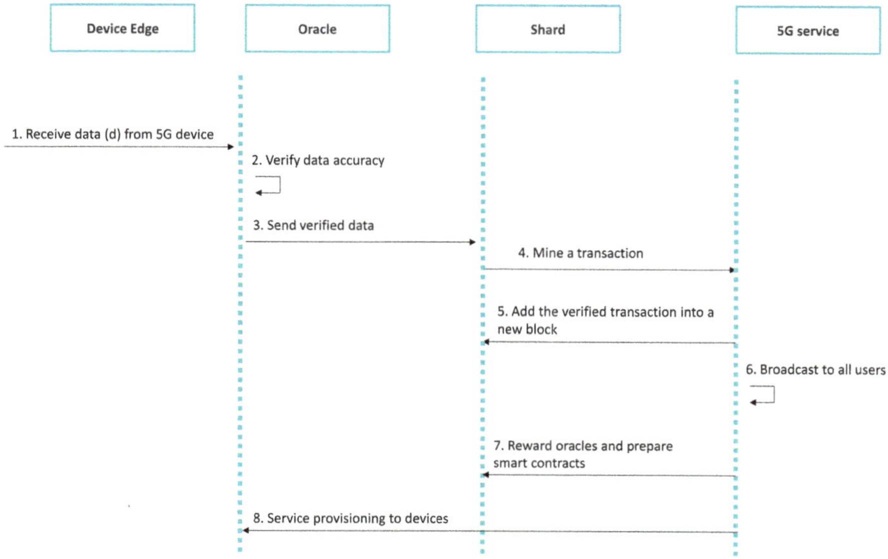

Fig. 1.6 A sequence diagram of utilizing oracles and shards with 5G services

reward amount that a node i used to verify that a transaction partition is equal to 0 or 1. In the case of a valid verification, the reward is as follows:

$$r_i = \frac{R(T_{i,j})}{Total} + \alpha_i, \quad (1.6)$$

where $R(T_{i,j})$ is the master cluster reward, $Total$ refers to the total number of the cluster devices participating in the verification process, and α_i node deposit. A node's reward is equal to his participating share in the transaction verification process. For instance, if a cluster has a reward equal to 1000, and the number of participant nodes is equal to 100, the reward is distributed equally, and each node will receive 100 as a stake. In the case of false verification, the node is penalized, and the reward is deducted from its deposit.

1.6 Challenges and Future Research Directions

Blockchain technology has the potential to improve 5G networks and to solve several issues ranging from security challenges to system management in a decentralized manner. Yet there are various challenges need further investigation before the deployment of blockchain in 5G networks. This section highlights and discusses some of the open challenges that may set back the merging of blockchain with 5G networks.

1.6.1 Scalability and Performance

The scalability of blockchain is measured by the rate at which transactions are added to the chain, in other words, the number of transactions per second (TPS). Currently, popular blockchain platforms, such as Bitcoins and Ethereum, can reach up to 14 TPS, whereas some private blockchains can reach up to 3000 TPS. Transaction speed is one of the major concerns for adopting blockchain technology. High TPS is essential for 5G networks to meet the requirements of low-latency and high communication rate. Straightforward integration of the current blockchain is unsuitable for 5G networks. Furthermore, blockchain has high throughputs because of the complexity of blockchain consensus. 5G networks require coordination between a large number of entities, services, and operations that cannot be achieved using blockchain. To solve these problems, further studies are required to improve the scalability of blockchain solutions to meet the requirement of 5G in a dynamic and heterogeneous manner and to involve a large number of transactions with low latency.

1.6.2 Standardization and Regulations

It is expected that various services and applications are going to be delivered by the 5G systems. However, standardizations are still challenging because of the interoperability, ubiquitously, and dynamicity of 5G communication systems and applications. In addition, blockchain just started gaining the attention of network operators. Consequently, to get a wide acceptance, network operators should work together to define a standardization of blockchain integration with 5G networks. Furthermore, blockchain smart contracts are de-standardized, deregulated, and unmodifiable; it is important to ensure the security of blockchain technology before its integration with 5G networks. Any falsified manipulation of blockchain smart contracts can severely damage the network and lead to serious consequences. For a large adoption of blockchain in the context of 5G networks, standardization, regulation, and governance must be enacted.

1.6.3 Resource Constraints

The blockchain was designed for an Internet scenario with powerful computers; it is computationally expensive and has significant overhead in terms of both bandwidth and storage capacity. These particular characteristics currently exclude the easy integration of the blockchain with 5G networks. Indeed, to take part in the blockchain, nodes (i.e., miners that can verify a transaction) need to have a high computing power to run a blockchain consensus [17]. In some situations, 5G nodes may be already taking part in an operation to provide a service and might not be able to have enough resources to run also the blockchain, which may lead to network degradation, bottleneck, and SPF issues. Due to resource limitations, a lightweight framework that can dynamically optimize resource usage among several nodes is required for the 5G networks. Consequently, resource provisioning for restrained nodes should be further investigated. Besides, IoT devices are resource-constrained. Incorporating blockchain technology with the IoT nodes is challenging. The high-computing-power storage capacity and overheads have put under question the feasibility of the integration of blockchain within the 5G.

1.6.4 Interoperability

Achieving seamless interoperability among blockchain and 5G networks remains a challenge. Indeed, several well-known blockchain platforms offer different features. Nevertheless, it is unclear how blockchain technology can be merged with 5G networks. Furthermore, 5G networks and beyond involve various new technologies such as NFV, SDN, MEC, massive MIMO, and full duplex [22–24]. Each of which works differently. These are key challenges that need to be addressed prior to the integration. Among the questions that are worth investigating are: how can we deploy blockchain in 5G networks? Who will be responsible for ensuring the interoperability among 5G services, operations, and users? Will blockchain going to be deployed as an overlay network?

1.6.5 Security

Even if blockchain is considered as a secure technology that establishes trust in 5G networks, it still suffers from various security issues, such as 51% attacks that occur when an attacker takes positing of 51% of all network computing power. This type of attack led to falsification, SPF, and data corruption. If one attacker takes the position of 51% computing power, then he can control all the blockchains and thus all services and operations running in the blockchain. Furthermore, blockchain smart contracts can cause security issues due to poorly written code. Safe deployment of blockchain technology with a 5G network should take into consideration the issue inherent to the blockchain. Furthermore, data privacy has become a major concern especially in the context of blockchain and 5G. For example, data related to a credit card used to pay for a service provided by the 5G network can be stored in the blockchain permanently and cannot be deleted. By design, private data should not be stored in the chain, but rather off-the chain and only have a pointer of that information. In this context, several researchers have suggested the use of on-the-chain and off-the-chain storage; on-the-chain stores will only have pointers, metadata, and hashes of the actual data stored off-the-chain in a secure database.

1.6.6 Infrastructure Costs

It is expected that 5G networks are going to increase energy consumption because of the increased amount of equipment. With ongoing efforts for energy-efficient communications and networking, green communications are going to be harder to achieve. The blockchain solution will be a hurdler for 5G networks because of its computationally intensive consensus algorithms. For example, the bitcoin blockchain is estimated to consume at a peak of more electricity than 159 countries. Furthermore, the use of blockchain platforms and cloud infrastructures that host blockchain nodes comes at a cost. In some situations, every blockchain transaction has fees. For example, in the Ethereum blockchain, a transaction can cost a gas unit that refers to the number of computation efforts required to verify a transaction. Transaction costs are relative to the complexity of the consensus running. Consequently, designing an energy-efficient blockchain consensus algorithm is a big hurdle for blockchain integration in the 5G networks.

1.7 Conclusion

Blockchain technology was originally intended for the cryptocurrency context; however, this technology has moved beyond its realm. Blockchain-based solutions have benefited from blockchain's features to improve their process. In this context, blockchain was proposed by several researchers as a solution to 5G inherent challenges. Indeed, several studies have shown the benefit of using blockchain solutions to meet some of the requirements of 5G networks such as security, transparency, decentralization, and immutability. Blockchain integration with 5G networks is expected to improve new technologies with a secure design concept, limited access control, and trust. Furthermore, blockchain smart contract, lightweight consensus, sharding concept, and oracle blockchain will enable several new 5G business models to benefit from a high data rate, secure communication, and reliable services. In this chapter, we have reviewed some blockchain integration opportunities that can empower 5G network services and operations. Based on this overview, we have proposed a scalable, secure, and lightweight blockchain architecture adapted to some 5G requirements. Furthermore, we have summarized some open challenges and provided some research directions that need further investigation for the safe deployment of blockchain in 5G networks.

Acknowledgments The authors would like to thank the Natural Sciences and Engineering Research Council of Canada, as well as FEDER and Grand Est Region in France, for the financial support of this research.

References

1. J. Navarro-Ortiz, P. Romero-Diaz, S. Sendra, P. Ameigeiras, J.J. Ramos-Munoz, J.M. Lopez-Soler, A survey on 5G usage scenarios and traffic models. IEEE Commun. Surv. Tutorials **22**(2), 905–929 (2020)
2. P.K. Agyapong, M. Iwamura, D. Staehle, W. Kiess, A. Benjebbour, "Design considerations for a 5G network architecture. IEEE Commun. Mag. **52**(11), 65–75 (2014)
3. A. Gupta R.K. Jha, A survey of 5G network: Architecture and emerging technologies. IEEE Access **3**, 1206–1232 (2015)
4. Z. Mlika, S. Cherkaoui, Massive access in beyond 5G IoT networks with NOMA: NP-hardness, competitiveness and learning. arXiv:2002.07957 [cs, math], Feb. 2020, arXiv: 2002.07957. [Online]. Available: http://arxiv.org/abs/2002.07957
5. X. Li, P. Jiang, T. Chen, X. Luo, Q. Wen, A survey on the security of blockchain systems. Futur. Gener. Comput. Syst. **107**, 841–853 (2020). [Online]. Available: https://www.sciencedirect.com/science/article/pii/S0167739X17318332
6. I.-C. Lin, T.-C. Liao, A survey of blockchain security issues and challenges. Int. J. Netw. Secur. **19**(5), 653–659 (2017)
7. A. Chaer, K. Salah, C. Lima, P.P. Ray, T. Sheltami, Blockchain for 5G: Opportunities and challenges, in *2019 IEEE GLOBECOM Workshops (GC Wkshps)* (2019), pp. 1–6
8. Z.A.E. Houda, A. Hafid, L. Khoukhi, Blockchain meets AMI: Towards secure advanced metering infrastructures, in *ICC 2020–2020 IEEE International Conference on Communications (ICC)* (2020), pp. 1–6

9. D.C. Nguyen, P.N. Pathirana, M. Ding, A. Seneviratne, Blockchain for 5G and beyond networks: a state of the art survey (2019). arXiv:1912.05062 [cs, eess, math], Dec. 2019, arXiv: 1912.05062. [Online]. Available: http://arxiv.org/abs/1912.05062
10. Z. Abou El Houda, A.S. Hafid, L. Khoukhi, Cochain-SC: An intra- and inter-domain DDoS mitigation scheme based on blockchain using SDN and smart contract. IEEE Access **7**, 98893–98907 (2019)
11. H. Moudoud, S. Cherkaoui, L. Khoukhi, An IoT blockchain architecture using oracles and smart contracts: the use-case of a food supply chain, in *2019 IEEE 30th Annual International Symposium on Personal, Indoor and Mobile Radio Communications (PIMRC)* (2019), pp. 1–6
12. B. Liu, X.L. Yu, S. Chen, X. Xu, L. Zhu, Blockchain based data integrity service framework for IoT data, in *2017 IEEE International Conference on Web Services (ICWS)* (2017)
13. Z. Abou El Houda, L. Khoukhi, A. Senhaji Hafid, Bringing intelligence to software defined networks: Mitigating DDoS attacks. IEEE Trans. Netw. Serv. Manag. **17**(4), 2523–2535 (2020)
14. X. Li, P. Jiang, T. Chen, X. Luo, Q. Wen, A survey on the security of blockchain systems. Futur. Gener. Comput. Syst. **107**, 841–853 (2020). [Online]. Available: https://www.sciencedirect.com/science/article/pii/S0167739X17318332
15. X. Xu, I. Weber, M. Staples, L. Zhu, J. Bosch, L. Bass, C. Pautasso, P. Rimba, A Taxonomy of blockchain-based systems for architecture design, in *2017 IEEE International Conference on Software Architecture (ICSA), April* (2017), pp. 243–252
16. N.M. Kumar, P.K. Mallick, Blockchain technology for security issues and challenges in IoT. Procedia Comput. Sci. **132**, 1815–1823 (2018). [Online]. Available: http://www.sciencedirect.com/science/article/pii/S187705091830872X
17. Z. Abou El Houda, A.S. Hafid, L. Khoukhi, Cochain-SC: An intra- and inter-domain DDoS mitigation scheme based on blockchain using SDN and smart contract. IEEE Access **7**, 98893–98907 (2019)
18. L.S. Sankar, M. Sindhu, M. Sethumadhavan, Survey of consensus protocols on blockchain applications, in *2017 4th International Conference on Advanced Computing and Communication Systems (ICACCS)* (IEEE, Coimbatore, 2017), pp. 1–5. [Online]. Available: http://ieeexplore.ieee.org/document/8014672/
19. Z.A. El Houda, L. Khoukhi, A. Hafid, Chainsecure - a scalable and proactive solution for protecting blockchain applications using SDN, in *2018 IEEE Global Communications Conference (GLOBECOM)* (2018), pp. 1–6
20. H. Moudoud, S. Cherkaoui, L. Khoukhi, Towards a scalable and trustworthy blockchain: IoT use case, in *ICC 2021–2021 IEEE International Conference on Communications (ICC)* (2021), pp. 1–6
21. H. Moudoud, L. Khoukhi and S. Cherkaoui, Prediction and detection of FDIA and DDoS attacks in 5G enabled IoT, in IEEE Network (2021) vol. 35, no. 2, pp. 194–201
22. Z.A.E. Houda, A. Hafid, L. Khoukhi, Blockchain-based reverse auction for V2V charging in smart grid environment, in *ICC 2021 - 2021 IEEE International Conference on Communications (ICC)* (2021), pp. 1–6
23. A. Filali, Z. Mlika, S. Cherkaoui, A. Kobbane, Preemptive SDN load balancing with machine learning for delay sensitive applications. IEEE Trans. Veh. Technol. **69**(12), 15947–15963 (2020)
24. Z.A. El Houda, A. Hafid, L. Khoukhi, Co-IoT: A collaborative DDoS mitigation scheme in IoT environment based on blockchain using SDN, in *2019 IEEE Global Communications Conference (GLOBECOM)* (2019), pp. 1–6

Chapter 2
Deep Learning Approach for Interference Mitigation in MIMO-FBMC/OQAM Systems

Abla Bedoui and Mohamed Et-tolba

2.1 Introduction

The next generations of wireless communication systems aim to come up with a wide range of heterogeneous services, which are fundamentally categorized into three use cases: enhanced mobile broadband (eMBB), ultra-reliable low-latency communication (URLLC), and millimeter-wave spectrum technologies [2, 21]. These services can be delivered simultaneously, while meeting the quality-of-service requirements including flexibility, availability, and system performance [13]. Therefore, the system architecture, particularly the physical layer, has to be designed in order to guarantee an agile exploitation of resources. For instance, to support several use cases in the fifth generation (5G) of wireless communications, a flexible new radio interface, called (5G NR), is developed by the 3rd group partnership project (3GPP). Among the novelties of 5G NR is the use of various multicarrier modulation techniques for data transmission [10]. Offset quadrature amplitude modulation-based filter bank multicarrier (FBMC/OQAM) is potentially considered as a candidate because of its improved spectral efficiency compared to the conventional multicarrier schemes. Indeed, FBMC/OQAM is robust against highly selective distortions while enabling a dynamic spectrum usage [12].

Despite the higher spectrum efficiency of FBMC/OQAM, its joint application with multiple-input and multiple-output (MIMO) is challenging. Therefore, the multiantenna configuration is not straightforward due to the intersymbol interference that has a considerable influence on the system performance. Several studies on the combination of MIMO and FBMC, considering interference mitigation, have been

A. Bedoui (✉) · M. Et-tolba
Department of Communication systems, National Institute of Posts and Telecommunications, Rabat, Morocco
e-mail: bedoui@inpt.ac.ma; ettolba@inpt.ac.ma

M. Ouaissa et al. (eds.), *Computational Intelligence in Recent Communication Networks*, EAI/Springer Innovations in Communication and Computing,
https://doi.org/10.1007/978-3-030-77185-0_2

carried out recently. In [22], the authors proposed a method that aims to update the space-frequency block code (SFBC) scheme to reduce FBMC/OQAM intrinsic interference power. A considerable amount of the interference is automatically removed only by performing a specific Alamouti decoding. The suggested scheme improves the signal-to-interference ratio (SIR) but not in the case of higher mapping orders as is considered. The method proposed in [15] uses a parallel equalization structure that can efficiently fight the high selectivity of the channel. This design employs parallel equalizers and analysis filter banks (AFB) working simultaneously on the received signal to reduce the destructive effect of the transmission environment. Xu et al. [19] suggested a more elegant solution to mitigate ISI in MIMO-FBMC/OQAM system based on a repeated block with the reverse arrangement of symbols. Moreover, they have developed a singular value decomposition (SVD)-based precoding solution to inhibit intercarrier interference (ICI) under frequency selective channel. This approach outperforms the Alamouti OFDM scheme. Unfortunately, this comes with an increase of the computational cost due to the use of SVD.

In this chapter, we aim to design an interference-free MIMO-FBMC/OQAM communication system using deep learning. This latter has emerged recently as a sophisticated tool that proved a significant effectiveness in wireless communication systems. It has been the subject of numerous studies. In [9], the authors proposed a supervised deep neural network (DNN) model using a specific training method. This method is based on the use of the received pilot symbols and the channel impulse response in order to provide the estimated channel impulse response as an output. The power of deep learning has also been shown in multicarrier modulation systems. For FBMC/OQAM, a deep-learning-based channel estimator and equalizer were investigated in [4]. In this approach, channel estimation, equalization, and demapping are treated as a black box approached by a DNN model. Besides, in [18], a Bayesian-learning-based doubly selective sparse channel estimation for millimeter-wave hybrid MIMO-FBMC/OQAM systems has been investigated in [10]. Additionally, an online BL-based Kalman filter (OBLKF) is designed for sparse channel tracking in doubly selective channel estimation techniques. On the other hand, the proposed techniques were seen to significantly outperform the conventional ones. Unfortunately, the interference mitigation was not taken into account in the previously cited work.

Hence, we proposed, we propose a novel interference-free detection method for MIMO-FBMC/OQAM using a deep neural network. We consider MIMO-FBMC/OQAM as a blind source separation (BSS) problem where the estimated sources (OQAM symbols) must be statistically independent. In this manner, we reduce the correlation and then the interference between the desired OQAM symbol, on a given subcarrier, and the ones carried by the other subcarriers. In the proposed BSS approach, the transmitted OQAM symbols are seen as the source signals, whereas the received signals represent the mixtures. Moreover, the blind prediction of the unmixing matrix is reduced to the optimization of the independent component analysis (ICA) rotation matrix. This optimization is mapped into a supervised regression where we treat the real and imaginary parts of the received mixtures separately. To find an appropriate mapping between the received mixtures (inputs) and their corresponding transmitted symbols (targets), we propose a suitable DNN architecture. The learning rule in the proposed DNN takes into account the statistical independence constraint, which is measured by Kullback–Leibler (KL) divergence [8].

The remainder of this chapter is organized as follows. In Sect. 2.2, we present the MIMO-FBMC/OQAM system model. In Sect. 2.3, we formulate the MIMO-FBMC/OQAM as a BSS problem. In Sect. 2.4, we present the proposed architecture for joint blind detection and interference mitigation using a DNN taking into account the statistical independence of the estimated sources (OQAM symbols). In Sect. 2.5, we present and analyze the simulation results. Finally, we conclude the chapter in Sect. 2.6.

2.2 MIMO-FBMC/OQAM System Model

FBMC/OQAM is a multicarrier modulation technique where the bandwidth is divided into M sub-bands spaced by Δ_f. It considers high-order quadrature amplitude modulation (QAM) whose output symbols, of duration T_s, are time-staggered to form real-valued OQAM symbols, each having a duration of $\frac{T_s}{2}$. The resulting symbol stream is assigned to the subcarriers, which individually apply a prototype filter to prevent interferences. The prototype filter must be designed in such a way that it increases the bandwidth efficiency and meets the orthogonality condition. Let us denote the transmit symbol at subcarrier position k and time position by $s_k(n)$. The discrete-time transmitted FBMC/OQAM signal can then be expressed as

$$x(m) = \sum_{k=0}^{K-1} \sum_{n} s_k(n) \underbrace{g(m - n\frac{K}{2})\beta_{k,m} e^{\frac{j2\pi km}{K}}}_{G_k(m)}, \tag{2.1}$$

where $g(m)$ is the impulse response of the prototype filter and $G_k(m)$ is the discrete-time impulse response of the synthesis filter corresponding to the kth subcarrier. Let L be the prototype filter length; the quantity $\beta_{k,m}$ is expressed as

$$\beta_{k,m} = e^{-j\frac{2\pi}{K}k(\frac{L-1}{2})}. \tag{2.2}$$

The FBMC/OQAM symbol stream is transmitted using the MIMO system depicted in Fig. 2.1. We assume that there are N_t transmit antennas and N_R receive antennas. We denote by $h_{(i,j)}$ the channel impulse response linking the ith transmit antenna and the jth receive antenna. Then, the demodulated OQAM symbol on the kth subcarrier at the jth antenna is written as

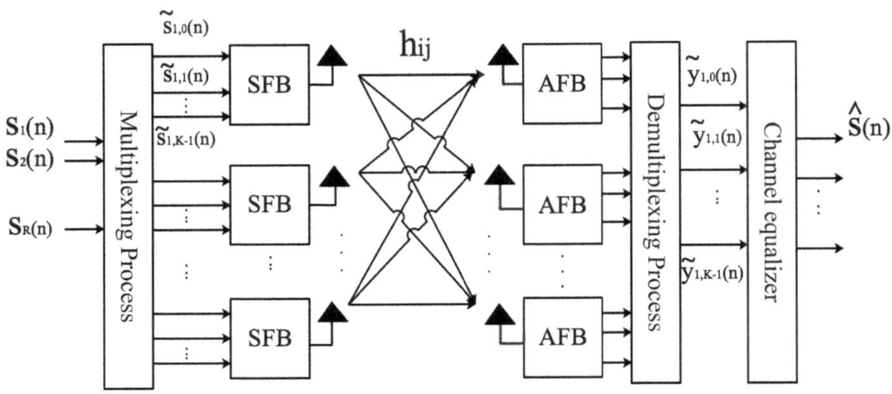

Fig. 2.1 MIMO-FBMC/OQAM discrete-time model

$$y_k^j(m) = \sum_{i=0}^{Nt} s_k^i(n) * F_{kq}^{ji} + \sum_{i=0}^{Nt} \sum_{q \neq k}^{K-1} s_q^i(n) * F_{kq}^{ji} + b(m), \tag{2.3}$$

where

$$F_{kq}^{ji} = G_k(m) * h_{i,j}(m) * G_q^*(-m). \tag{2.4}$$

The first term on the right-hand side of Eq. (2.4) stands for the desired transmitted signal, while the second term is the interference expression and $b(m)$ is the additive white Gaussian noise. On the other hand, the matched filter is represented by $\boldsymbol{G}_q^*(-m)$, which is the conjugate of the basis pulse.

The first term on the right-hand side of Eq. (2.4) stands for the desired transmitted signal, the second term is the intersymbol interference, and $b(m)$ is the additive white Gaussian noise. The quantity $\boldsymbol{G}_q^*(-m)$ in (2.4) represents a matched filter, which is the conjugate of the basis pulse. In order to have maximum spectral efficiency and eliminate the imaginary interference, it is mandatory to switch to a less strict real orthogonality condition, in such a way that only real-valued symbols can be transmitted [14]. In the case of single-input single-output (SISO) transmission systems, a common way is to reduce the time spacing as well as the frequency spacing that causes interference that is somehow fetched to the purely imaginary domain. Hence, the fact of taking the real part leads to the mitigation of imaginary interference and allows for simple detection process. Unfortunately, this limits the applicability of certain MIMO techniques such as space–time block codes. Moreover, the presence of interference between streams and antennas in MIMO needs a more sophisticated method for interference mitigation. Besides, in the case of using a good time–frequency localization prototype filter such as PHYDYAS, the intrinsic interference is mainly upcoming from the neighboring time and frequency positions [12].

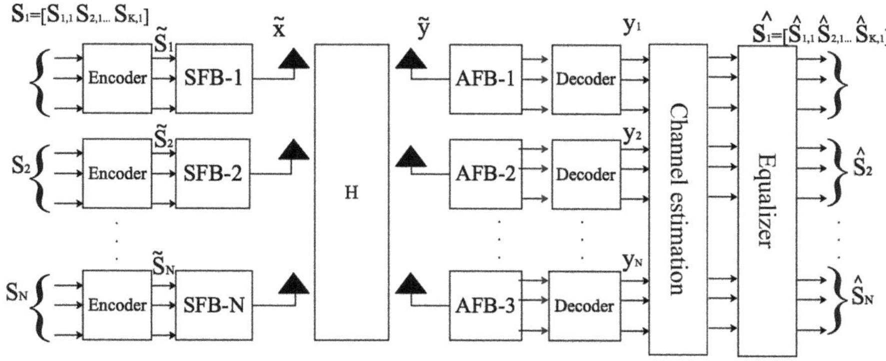

Fig. 2.2 MIMO-FBMC/OQAM matricial model

For ease of exposition, we observe a FBMC/OQAM symbol duration and establish the system model in matrix notation as depicted in Fig. 2.2. Let $\boldsymbol{s} \in \mathbb{C}^{1\times KN}$ be the vector containing the concatenation of the OQAM symbols carried by K subcarriers and will be sent from the N_T antennas,

$$\boldsymbol{s} = \left[\boldsymbol{s}_1\ \boldsymbol{s}_2\ \ldots\ \boldsymbol{s}_N\right] = [s_{1,1}s_{2,1}...s_{K,1}...s_{1,N}...s_{K,N}]. \tag{2.5}$$

As stated above, the main problem of FBMC/OQAM is that we can only transmit real-valued data. Thus, the possibility of applying MIMO with FBMC/OQAM cannot be straightforwardly done as in OFDM. This limitation has been resolved in [14] by spreading symbols in frequency using a unitary coding matrix $\boldsymbol{C} \in \mathbb{C}^{2NK\times NK}$.

However, the use of a huge encoding matrix $\boldsymbol{C}$ may introduce a high computational complexity at the receiver while calculating $\boldsymbol{C}^{-1}$. Thus, a block-wise encoding method has been proposed in [14], where the N transmitted symbols are precoded separately so as to decrease the size of the encoding matrix. Hence, the $\tilde{\boldsymbol{s}} \in \mathbb{C}^{2KN\times 1}$ precoded transmitted signal can be expressed as follows:

$$[\tilde{\boldsymbol{s}}_1\tilde{\boldsymbol{s}}_2\ldots\tilde{\boldsymbol{s}}_N] = [\boldsymbol{C}\boldsymbol{C}\ldots\boldsymbol{C}]\begin{bmatrix}\boldsymbol{s}_1\\ \boldsymbol{s}_2\\ \vdots\\ \boldsymbol{s}_N\end{bmatrix}, \tag{2.6}$$

and let $\boldsymbol{G}_i \in \mathbb{C}^{L\times 2K}$, where $i = \{1, 2, ...N\}$ be the matrices representing the synthesis filters for each block component and $\boldsymbol{C} \in \mathbb{C}^{2K\times K}$ be the encoding matrix. Accordingly, the block of the transmitted FBMC/OQAM symbols $\tilde{\boldsymbol{x}} \in \mathbb{C}^{LN\times 1}$ is expressed as

$$\tilde{\boldsymbol{x}} = [\tilde{\boldsymbol{x}}_1 \tilde{\boldsymbol{x}}_2 \ldots \tilde{\boldsymbol{x}}_N] = [\boldsymbol{G}_1\boldsymbol{C} \quad \boldsymbol{G}_2\boldsymbol{C} \ldots \boldsymbol{G}_N\boldsymbol{C}]\boldsymbol{H} \begin{bmatrix} s_1 \\ s_2 \\ \vdots \\ s_N \end{bmatrix} + \boldsymbol{b}, \tag{2.7}$$

where $\boldsymbol{H} \in \mathbb{C}^{LN \times LN}$ is the channel matrix. The vector $\boldsymbol{b} \in \mathbb{C}^{LN \times 1}$ represents an additive white Gaussian noise (AWGN) vector with mean zero and covariance matrix $\boldsymbol{C_b} = \sigma_b^2 \boldsymbol{I}$. The quantities σ_b^2 and $\boldsymbol{I}$ are the noise variance and the identity matrix, respectively. After being transmitted over the transmission channel, the received signal $\tilde{\boldsymbol{x}}$ is FBMC-demodulated and then decoded (despreaded). The output $\boldsymbol{y_t}$ corresponding to the tth block is written as

$$\boldsymbol{y}_t = \boldsymbol{C}^H \boldsymbol{G}_t^H \tilde{\boldsymbol{x}} + \boldsymbol{C}^H \boldsymbol{G}_t^H \boldsymbol{b}_t, \tag{2.8}$$

where subscript H refers to the transpose conjugate operator.

The combination of (2.21) and (2.9) leads to

$$\boldsymbol{y}_t = \boldsymbol{s}_t + \sum_{r=1 r \neq t}^{N} \boldsymbol{C}^H \boldsymbol{G}_t^H \boldsymbol{G}_r \boldsymbol{C} \tilde{\boldsymbol{x}}_r + \boldsymbol{C}^H \boldsymbol{G}_t^H \boldsymbol{b}_t, \tag{2.9}$$

where the second term in the right-hand side represents the inter-block interference and $\boldsymbol{b}_t$ is the AWGN vector corresponding to the tth block. The intercarrier interference can be highlighted considering the received symbol, at the kth subcarrier and code position c in the tth block, which is expressed as

$$\begin{aligned} y_{k,c,t} &= c_{k,c}^H \boldsymbol{G}_t^H \boldsymbol{H} \tilde{\boldsymbol{x}}_t + b_{k,t} \\ &= s_{k,c,t} + \sum_{q=1, q \neq k}^{K} c_{k,c}^H \boldsymbol{G}_t^H \boldsymbol{H}_{\boldsymbol{k},\boldsymbol{q}} \boldsymbol{G}_t \tilde{\boldsymbol{x}}_t + b_{k,t}, \end{aligned} \tag{2.10}$$

where the second term on the right-hand side represents the intercarrier interference inside the tth block.

The approach we propose for interference cancellation is based on unsupervised deep neural network in which the learning is done with a modified loss function depending on the statistical independence constraint function as detailed in the next section.

2.3 Problem Formulation

As mentioned previously, we deal with interference mitigation in MIMO-FBMC/OQAM systems using BSS formulation. Indeed, the received FBMC/OQAM

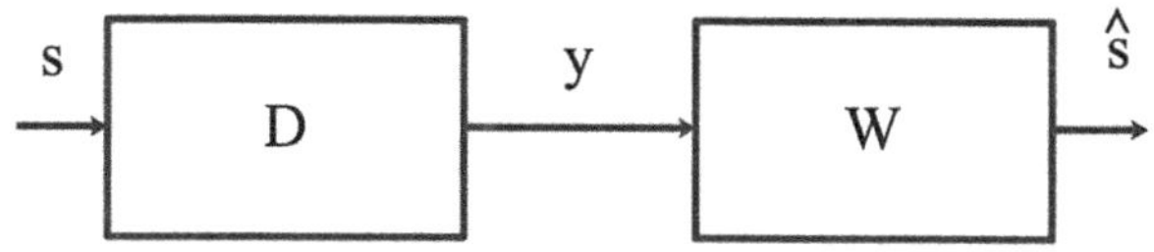

Fig. 2.3 Blind source separation equivalent model

symbols (observations) are mixtures of the transmitted ones (sources) where the transmission matrix $\boldsymbol{D}$ is considered as the mixing matrix as depicted in Fig. 2.3 and presented as follows:

$$\boldsymbol{y} = \boldsymbol{D}\boldsymbol{s} + \boldsymbol{b}, \tag{2.11}$$

where

$$\boldsymbol{D} = [\boldsymbol{G}_1\boldsymbol{C} \quad \boldsymbol{G}_2\boldsymbol{C} \ldots \boldsymbol{G}_N\boldsymbol{C}]\boldsymbol{H}\begin{bmatrix} \boldsymbol{G}_1^H\boldsymbol{C}^H \\ \boldsymbol{G}_2^H\boldsymbol{C}^H \\ \vdots \\ \boldsymbol{G}_N^H\boldsymbol{C}^H \end{bmatrix}. \tag{2.12}$$

Let $\boldsymbol{W}$ be the blind equalization matrix that represents the unmixing matrix, which is established in such a way that the recovered sources are statistically independent. In statistics, when two random variables are statistically independent, it results in a certain no-correlation, i.e., the absence of covariance. Furthermore, for a specific element, it is well known that interference from other elements is a correlation between the desired element and other elements. Consequently, the high-order statistical independence between the received vector elements has two major advantages, namely the absence of the inter-block and intercarrier interference. In this chapter, the number of sources is not the number of MIMO transmitter antennas since the source signals are not the signals corresponding to MIMO antennas as is considered in the previous works. Besides, the sources are the OQAM symbols before being coded and FBMC modulated as depicted in Fig. 2.5. Hence, the number of sources is the same as the number of mixtures, which leads to a determined mixture. In this case, the source separation problem can be performed by computing the separating matrix so that the estimated source signals are as independent as possible. Consequently, this matrix operates as a linear spatial equalizer that recovers the transmitted OQAM symbols and simultaneously attenuates the interference.

Before applying any separating method of the convolutive mixtures, we shall consider three assumptions that are typically made on the source separation problem [11, 16]:

Recall that the desired components are extracted from a mixture of signals using the independent component analysis (ICA). This aims to find a linear representation of the observed data as non-Gaussian and mutually independent latent variables, also called independent components [6]. Let $\boldsymbol{W}$ be the unmixing matrix; then, the estimated source vector $\hat{\boldsymbol{s}}$ is expressed as

$$\hat{s} = \boldsymbol{W}\boldsymbol{y}. \tag{2.13}$$

Using matrix decomposition, the mixing matrix can be written as a product of the three matrices via some linear operations. An orthogonal matrix $\boldsymbol{V}$ is made up of the eigenvectors of $\boldsymbol{D}\boldsymbol{D}^T$. A second orthogonal matrix $\boldsymbol{U}$ is composed of eigenvectors of $\boldsymbol{D}\boldsymbol{D}^T$, and a pseudo-diagonal matrix $\Sigma = diag(\lambda_1, ..., \lambda_k, ..., \lambda_R)$, where λ_k are the eigenvalues.

$$\boldsymbol{D} = \boldsymbol{U}\boldsymbol{\Sigma}\boldsymbol{V}^T. \tag{2.14}$$

From the fact that the unmixing matrix is the inverse of the mixing one, we have

$$\boldsymbol{W} = \boldsymbol{D}^{-1} = (\boldsymbol{U}\boldsymbol{\Sigma}\boldsymbol{V}^T)^{-1} = \boldsymbol{V}\boldsymbol{D}^{-\frac{1}{2}}\boldsymbol{E}^T, \tag{2.15}$$

where $\boldsymbol{D}$ and $\boldsymbol{E}$ are, respectively, the eigenvalues and eigenvectors of the covariance of the received mixture. Substituting equation (2.15) into (2.13), we find

$$\hat{s} = \boldsymbol{V}\boldsymbol{y}_{\boldsymbol{W}}, \tag{2.16}$$

where $\boldsymbol{y}_{\boldsymbol{W}}$ is the whitened vector, which is obtained by a linear transformation of the observed vector $\boldsymbol{y}$. Note that the components of the whitened vector are uncorrelated with a unit variance vector, i.e., $\boldsymbol{y}_{\boldsymbol{W}}$ is white. Equivalently, the covariance matrix of $\boldsymbol{y}_{\boldsymbol{W}}$ equals the identity matrix:

$$E\{\boldsymbol{y}_{\boldsymbol{W}}\boldsymbol{y}_{\boldsymbol{W}}^T\} = \boldsymbol{I} \tag{2.17}$$

$$\boldsymbol{y}_{\boldsymbol{W}} = \boldsymbol{y}\boldsymbol{D}^{-\frac{1}{2}}\boldsymbol{E}^T. \tag{2.18}$$

As a result, the whitening is realized by two linear operations based on $\boldsymbol{D}$ and $\boldsymbol{E}$, where the received mixture is first projected on the principal components $\boldsymbol{D}^T$ and then scaled in such a way that every direction has unit variance $\boldsymbol{D}^{-\frac{1}{2}}\boldsymbol{E}^T$.

The problem of ICA is then reduced to the optimization of the rotation matrix $\hat{\boldsymbol{V}}$. This optimization shall require higher-order independence in order to equivalently eliminate the interference between the source signals. Several algorithms have been treated in the literature. They are mainly based on minimizing a contrast function that measures statistical independence of the received FBMC/OQAM symbols, respectively.

$$\hat{\boldsymbol{V}} = \underset{\boldsymbol{V}}{\text{argmax/min}}\left(\sum_k f(\boldsymbol{V}\boldsymbol{y}_{\boldsymbol{W}})\right). \tag{2.19}$$

There are numerous contrast functions exist for statistical independence measure including the concepts of kurtosis, negentropy, and mutual information [5]. In this chapter, we are interested in the mutual information contrast function. In detection problems, the Kullback–Leibler KL divergence as a special case of mutual information is a fundamental measure of the distance between two probability distributions [6]. In the next section, we show how to use KL divergence to estimate the rotation matrix using a DNN.

2.4 Deep Neural Network for Blind Detection and Interference Mitigation in MIMO-FBMC/OQAM Systems

Figure 2.4 depicts the proposed architecture, where a deep neural network is employed to estimate the original FBMC/OQAM transmitted symbols with the aid of BSS. Therefore, in order to totally eliminate the intersymbol interference, the estimation problem is mapped into the optimization of the rotation matrix of the BSS model as explained in the previous section. This regression is resolved by exploiting the high efficiency of the DNNs. Moreover, the weights of the DNN are adjusted while taking into consideration the constraint of statistical independence of the received symbols. The idea is to offline train the DNN using the elements of the transmitted and received symbols (OQAM) as input and target values, respectively. Then, the trained DNN is used to estimate the transmitted symbols during real-time transmission. The received symbols are first whitened before being used by the DNN to perform the blind estimation of the transmitted OQAM symbols in such a way that the interference is eliminated. In the following, we describe the architecture of the used DNN [17, 18]. We present the training procedure and develop a novel learning rule for the proposed model.

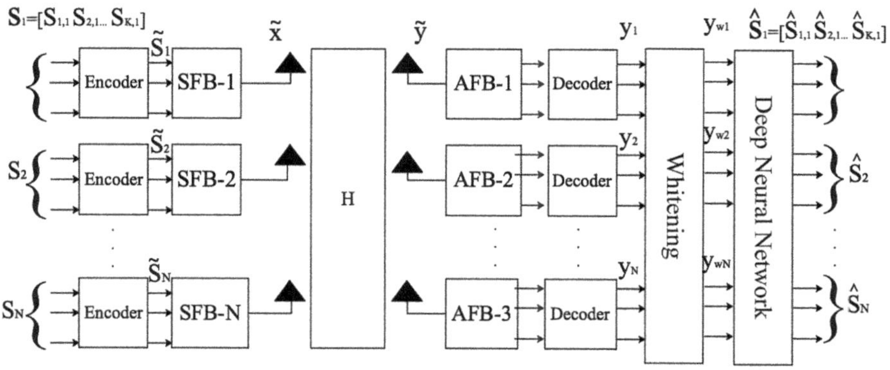

Fig. 2.4 Proposed deep learning detection approach for MIMO-FBMC/OQAM system

2.4.1 Data Set

The proposed neural network structure is presented in Fig. 2.5. It is composed of five fully connected layers, namely the input layer, three hidden layers, and the output layer. The training of this DNN for rotation matrix estimation is mainly based on a set of training samples $(\boldsymbol{Y}_{tr}, \boldsymbol{S}_{tr})$ including the whitened mixed signals as an input vector $\boldsymbol{Y}_{tr}$ and the source signals as a target vector $\boldsymbol{S}_{tr}$ formed by the transmitted block of NK OQAM symbols. This training data set is allied with the learning rule and used later by the DNN, so it can make an accurate prediction when a newly received OQAM symbol is presented.

Generating the training samples can be effectively done using many practical channel models such as ITU-Vehicular A (VA) and ITU-Pedestrian A (PA) profiles. Hereafter, we explain the steps followed to construct a data set for the DNN training:

1. We consider a random vector s_{tr} of N transmitted symbols and stock it as a target sequence for the DNN training.

$$\boldsymbol{S}_{tr} = \begin{bmatrix} s_{tr1} \\ s_{tr2} \\ \vdots \\ s_{trN} \end{bmatrix}. \tag{2.20}$$

2. The OQAM symbols are passed by the encoder and FBMC modulator and then convolved with the time-varying generated channel and added with AWG noise.

$$\tilde{\boldsymbol{x}}_{tr} = [\boldsymbol{G}_1\boldsymbol{C} \quad \boldsymbol{G}_2\boldsymbol{C} \ldots \boldsymbol{G}_N\boldsymbol{C}]\boldsymbol{H}\boldsymbol{S}_{tr} + \boldsymbol{b}_{tr}. \tag{2.21}$$

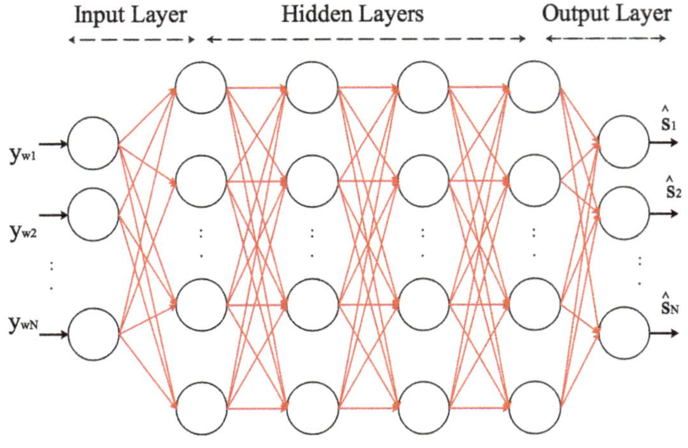

Fig. 2.5 Deep neural network structure for MIMO-FBMC/OQAM system

3. The received sequences (channel outputs) are FBMC-demodulated and decoded. The trained sequence $\tilde{\boldsymbol{x}}_{tr}$ is delivered to the FBMC demodulator and then to the decoder that outputs the sequence $\boldsymbol{y}$ given by

$$\boldsymbol{y}_{tr} = [\boldsymbol{G}_1\boldsymbol{C} \quad \boldsymbol{G}_2\boldsymbol{C} \ldots \boldsymbol{G}_N\boldsymbol{C}]\boldsymbol{H} \begin{bmatrix} \boldsymbol{G}_1^H\boldsymbol{C}^H \\ \boldsymbol{G}_2^H\boldsymbol{C}^H \\ \vdots \\ \boldsymbol{G}_N^H\boldsymbol{C}^H \end{bmatrix} \boldsymbol{S}_{tr} + \boldsymbol{b}_{tr}. \tag{2.22}$$

Also can be written as

$$\boldsymbol{y}_{tr} = \boldsymbol{D}_{tr}\boldsymbol{S}_{tr} + \boldsymbol{b}_{tr}, \tag{2.23}$$

where $\boldsymbol{D}_{tr}$ represents the training mixing matrix.

4. We construct a new vector composed of the concatenation of the real and imaginary parts of $\boldsymbol{y}_{tr}$,

$$\boldsymbol{Y}_{tr} = \begin{pmatrix} \Re\{\boldsymbol{y}_{tr}\} \\ \Im\{\boldsymbol{y}_{tr}\} \\ . \end{pmatrix} \tag{2.24}$$

5. The $\boldsymbol{Y}_{tr}$ vectors are headed to whitening operation using a predefined MATLAB function and then concatenated and stocked as an input sequence. Hence, the training set for one iteration can be presented as

$$(\boldsymbol{Y}_{tr}, \boldsymbol{S}_{tr}) = \left(\begin{bmatrix} \boldsymbol{Y}_{1tr} \\ \boldsymbol{Y}_{2tr} \\ \vdots \\ \boldsymbol{Y}_{Ntr} \end{bmatrix}, \begin{bmatrix} \boldsymbol{S}_{1tr} \\ \boldsymbol{S}_{2tr} \\ \vdots \\ \boldsymbol{S}_{Ntr} \end{bmatrix} \right). \tag{2.25}$$

These steps are repeated, and the training sets are concatenated until we have a huge input and target arrays.

2.4.2 Learning Rule

Learning to estimate the rotation matrix is the problem of optimizing the mapping between the whitened received mixture and the corresponding transmitted source. First, the input symbols are passed from the input layer to the hidden layers (composed of neurons) where different transformations are applied. In this chapter, we have three fully connected hidden layers; each one takes values from the previous layer and applies multiple operations such as: addition, multiplication, and activation operations before passing these values to the next layer using one of the activation functions used in deep learning such as ReLU, SeLU, and TanH [7]. The resulting values pass to the next layer at the end of this processing. At the output layer, the input from the last hidden layer is received and used to produce the final result. Typically, in the case of regression, a linear activation function is used in the output layer. Mathematically, the output of the nth neuron in the Lth hidden layer z_n^L can be expressed as

$$z_n^L = \sigma\left(\sum_m \theta_{n,m}^L\left(\cdots\left(\sigma\left(\sum_j \theta_{k,j}^2\left(\sigma\left(\sum_i \theta_{j,i}^1 z_i + b_j^1\right) + b_k^2\right)\right)\cdots\right)_m + b_n^L\right)\right)_n, \tag{2.26}$$

where $\sigma(.)$ is the activation function and $\theta_{n,m}^L$ is the weight from the mth neuron in the $(L-1)$th layer to the nth neuron in the Lth layer [1].

Our DNN is composed of three hidden layers with Leaky ReLu activation function similar to that used in [7],

$$\sigma(z) = max(0, z) + \alpha.min(0, z), \quad \alpha = 0.2. \tag{2.27}$$

During the learning process, the goal is to update the weights $\boldsymbol{\theta}$ of the DNN so as to minimize the difference between the predicted values and the corresponding targeted ones. This can be realized by Adam optimizer where the optimal value of $\boldsymbol{\theta}$ that minimizes the loss function L_1 is given by

$$L_1 = \frac{1}{K_I}\sum_{k_i=0}^{K_I-1}(S_{trk_i} - \theta_{k_i}Y_{trk_i}), \tag{2.28}$$

where K_I is the number of the neurons in the input layer and s_{k_i} is the k_ith targeted output value of DNN [7]. The vector of optimal weights is expressed as

$$\boldsymbol{\theta}_{opt} = \underset{\boldsymbol{\theta}}{\text{argmin}}(L_1). \tag{2.29}$$

In this chapter, the optimization of the rotation matrix is conceived under the condition that the components of the recovered signals are as independent as possible in order to guarantee interference elimination. To do so, we introduce the constraint of statistical independence on the learning rule. The one way to enforce a constraint on the output of the hidden layer is to add a regularization term in the expression of the loss function [20]. As discussed in the previous section, KL divergence is an adapted function for measuring how two variables are statistically independent. Let p_i be the probability distribution of the ith output of the last hidden layer; the KL divergence takes a large value when the distributions p_{k_o} and p_i diverge from each other and the value zero in the opposite case. Thus, the independence regularization can be written as

$$\Theta_{independence} = \frac{1}{\sum_{k_o=1}^{K_O} KL(p_{k_o}||p_i)}, \tag{2.30}$$

where K_O is the number of the outputs of the last hidden layer.

Eventually, minimizing the loss function L_2 leads to maximizing the KL divergence and then the statistical independence between the estimated sources. Hence, the modified loss function is written as

$$L_2 = \frac{1}{K_I} \sum_{k_i=0}^{K_I-1} (s_{trk_i} - \theta_{k_i} y_{trk_i}) + \Theta_{independence}. \tag{2.31}$$

The fact of maximizing the KL divergence leads to minimizing the correlation between the block components. This also means minimizing the interference between these components.

2.5 Simulation Results

We conducted computer simulations to compare the proposed deep-learning-based blind source separation for MIMO-FBMC/OQAM communication systems with conventional methods. In these simulations, we measured the block error rate (BLER), the neural network loss, and root mean square error (RMSE) as a function of SNR. These measurements are done using the Monte Carlo simulation method in both the vehicular and pedestrian channel profiles with the simulation parameters detailed in Table 2.1.

2.5.1 Deep Neural Network Performance: RMSE and Loss

Above all, it is essential to present the performance of the deployed DNN in this chapter. Hence, we performed computer simulations using the practical channel models VA and PA to construct a database of 5×10^7 training samples. Furthermore,

Table 2.1 Simulation parameters

Parameters	Values
Modulation	16-QAM
Subcarrier number	12
N_R	2
N_T	2
Subcarrier spacing	15 kHz
O_F	8
L_{ch}	8
Velocity	$V_{VA} = 200$ km/h, $V_{PA} = 2$ km/h

we employed more than 100 training epochs within the training rate of 10^{-3}. To measure DNN performance for rotation matrix prediction, we utilize two main metrics: RMSE and loss.

RMSE is known as the standard deviation of the residuals. It measures how well the regression algorithm fits the data point. Thus, Fig. 2.6 shows the training and validation RMSE as a function of the number of training iterations when SNR=10dB in the case of the PA channel profile. As can be seen, the network achieves its best performance at the 9th iteration where the RMSE values become constant. Additionally, it can be noticed that the proposed learning rule L_2 outperforms the conventional learning rule L_1.

One other metric to evaluate the performance of DNN is the loss. It is usually the residual summation of square errors (between the target outputs and the actual outputs) corresponding to each sample in the training and validation sets. In other words, loss tells how well is the model behavior during the training and testing process. In Fig. 2.6, we present the loss of the DNN used in this chapter for SNR = 10 dB. As the RMSE, the loss keeps the same value after the 9th iteration. Naturally, the training process using the L_2 learning rule provides a good fitting performance as compared to the learning rule L_1 since the loss value is lower for the proposed model. On the other hand, we describe it in Table 2.2 the number of training iterations attained to meet the validation criterion during the training operation as a function of SNR for the proposed learning rule. We notice that in the case of small SNR, a huge number of iterations are needed; besides, the values of loss and RMSE are still huge. This no accuracy can be explained by the high selectivity of the channel, but it can be solved by increasing the size of the training data set or the number of training iterations.

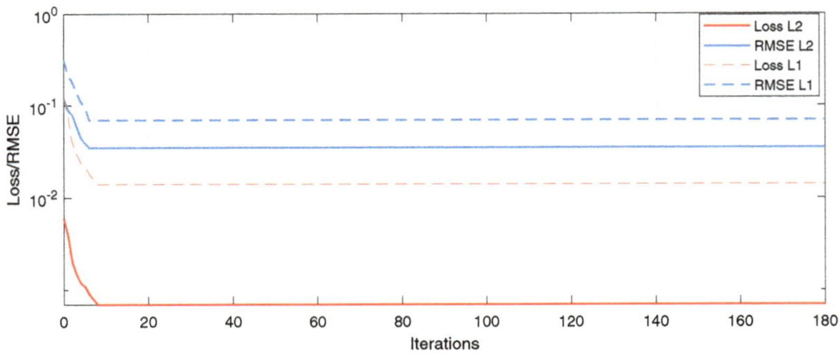

Fig. 2.6 Training RMSE and loss progress in terms of the number of iterations for SNR = 10 dB

Table 2.2 The number of training iterations attained to meet the validation criterion during the training process as a function of SNR

SNR	Iterations	RMSE	Loss	Max epoch
1	4200	3.73	7.30	109
4	1230	0.83	0.043	86
7	787	0.12	0.001	52
10	210	0.035	0.7×10^{-3}	7
13	132	0.016	0.1×10^{-3}	5

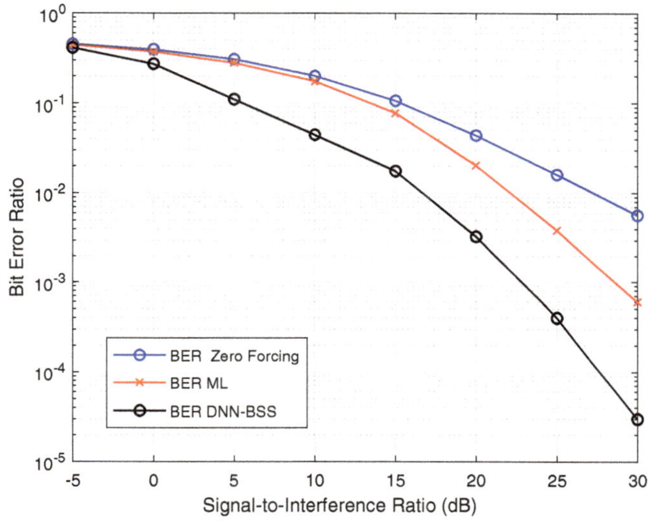

Fig. 2.7 BER performance of the proposed interference-free receiver for MIMO-FBMC/OQAM system in the case of doubly flat channel

2.5.2 *Bit Error Rate*

We also run a computer simulation to evaluate the BER of the proposed method and compared it to that obtained with the conventional detection methods: Maximum likelihood (ML) and zero forcing (ZF) for MIMO-FBMC/OQAM system. In Fig. 2.7, we present the BER as a function of SIR in the case of a doubly flat channel model. We can notice the gain of the proposed approach as compared to ML and ZF.

On the other hand, the BER results are represented for the VA channel profile in Fig. 2.8. As it is shown, the proposed method delivers an important gain as compared to ML and ZF channel equalization for MIMO-FBMC/OQAM that remains non-performant when the channel is severely doubly selective. Clearly, this remarkable gain in terms of SIR proves the interference mitigation ensured by the proposed system. Thus, the proposed blind detection method fights the high degradation of the vehicular transmission scenario. Besides, for the PA channel model, the BER values of ZF and ML are still not accurate even for less severe propagation conditions as depicted in Fig. 2.9. Thus, the proposed approach outperforms the conventional MIMO-FBMC/OQAM system. Accordingly, a considerable gap is in

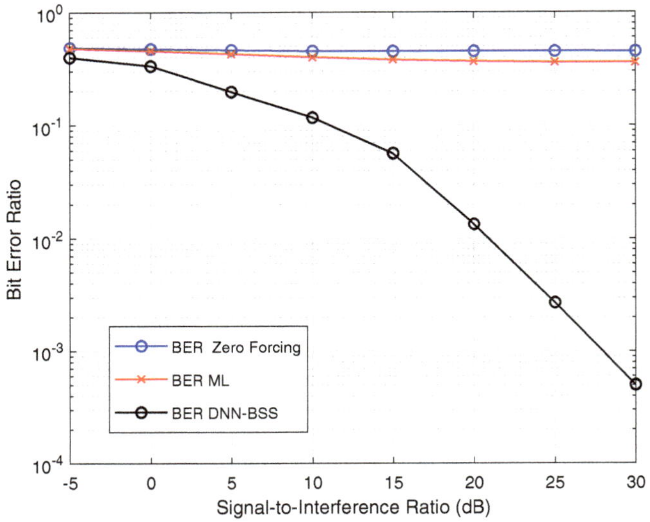

Fig. 2.8 BER performance of the proposed interference-free receiver for MIMO-FBMC/OQAM system in the case of VA channel profile

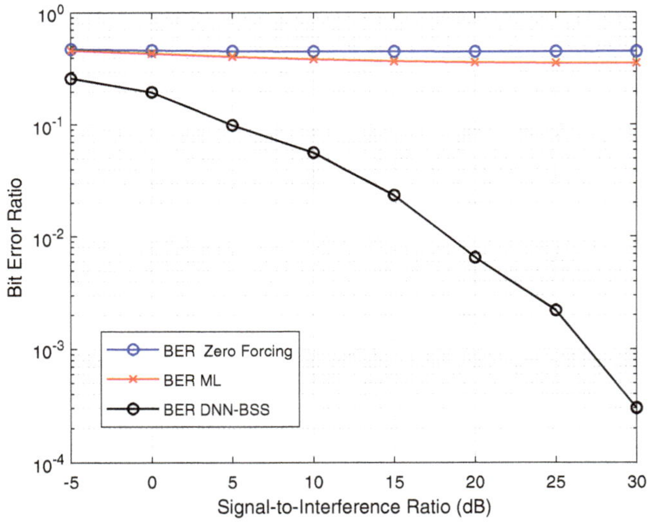

Fig. 2.9 BER performance of the proposed interference-free receiver for MIMO-FBMC/OQAM system in the case of PA channel profile

BER performance between the proposed method and the case of the conventional methods.

2.6 Conclusion

In summary, this chapter argued that the MIMO-FBMC/OQAM blind detection problem can be modeled as a BSS problem. Accordingly, the statistical independence constraint imposed by the BSS model leads to total interference elimination in the received mixtures (OQAM symbols). Then, we have proved that our BSS problem is transformed into the estimation of the ICA rotation matrix, while only the mixtures of MIMO-FBMC/OQAM received symbols are available at the receiver side. Thus, we have developed a supervised DNN to estimate the rotation matrix by optimizing the mapping between the received mixtures (inputs) and their corresponding transmitted sources (targets). On the other hand, we adapted our DNN to the BSS problem by enforcing the statistical independence presented as Kullback–Leibler (KL) divergence to our learning rule.

References

1. C.C. Aggarwal *Neural Networks and Deep Learning* (Springer, Cham, 2018)
2. I.F. Akyildiz, A. Kak, S. Nie, 6G and beyond: the future of wireless communications systems. IEEE Access **8**, 133995–134030 (2020)
3. M. Bellanger, PHYDYAS Project FP7 (2008). http://www.ict-phydyas.org/
4. X. Cheng, D. Liu, C. Wang, S. Yan, Z. Zhu, Deep learning-based channel estimation and equalization scheme for FBMC/OQAM systems. IEEE Wirel. Commun. Lett. **8**, 881–884 (2019). https://doi.org/10.1109/LWC.2019.2898437
5. J-T. Chien, *Source Separation and Machine Learning* (Elsevier, New York, 2019)
6. P. Comon, Independent component analysis—a new concept? Signal Process. **36**, 287–314 (1994)
7. I. Goodfellow, Y. Bengio, A. Courville, *Deep Learning* (MIT Press, Cambridge, 2016)
8. J. Harmouche, C. Delpha, D. Diallo, Incipient fault detection and diagnosis based on Kullback–Leibler divergence using principal component analysis: Part I. Signal Process. **94**, 278–287 (2014)
9. R. Jiang, X. Wang, S. Cao, J. Zhao, X. Li, Deep neural networks for channel estimation in underwater acoustic OFDM systems. IEEE Access **7**, 23579–23594 (2019). https://doi.org/10.1109/ACCESS.2019.2899990
10. S. Kumar, P. Singh, Filter bank multicarrier modulation schemes for visible light communication. Wireless Pers. Commun. **113**, 2709–2722 (2020)
11. B. Laheld, J.-F. Cardoso, Adaptive source separation with uniform performance. EURASIP **2**, 183–186 (1994)
12. J. Nadal, C.A. Nour, A. Baghdadi, Design and evaluation of a novel short prototype filter for FBMC/OQAM modulation. IEEE Access **6**, 19610–19625 (2018)
13. Y.L. Lee, D. Qin, L.-C. Wang, G.H. Sim, 6G massive radio access networks: key applications, requirements and challenges. IEEE Open J. Veh. Technol. **2**, 54–66 (2021)
14. R. Nissel, J. Blumenstein, M. Rupp, Block frequency spreading: A method for low-complexity MIMO in FBMC-OQAM, in *IEEE 18th International Workshop on Signal Processing Advances in Wireless Communications (SPAWC)*, pp. 1–5 (2017). https://doi.org/10.1109/SPAWC.2017.8227812

15. F. Rottenberg, X. Mestre, D. Petrov, F. Horlin, J. Louveaux, Parallel equalization structure for MIMO FBMC-OQAM systems under strong time and frequency selectivity. IEEE Trans. Signal Process. **65**, 4454–4467 (2017)
16. H. Sawada, N. Ono, H. Kameoka, D. Kitamura, H. Saruwatari, A review of blind source separation methods: Two converging routes to ILRMA originating from ICA and NMF. APSIPA Trans. Signal Inf. Process. **8**, E12. https://doi.org/10.1017/ATSIP.2019.5
17. B. Seo, D. Sim, T. Lee, C. Lee, Efficient time synchronization method with adaptive resource configuration for FBMC systems. IEEE Trans. Commun. **68**, 5563–5574 (2020)
18. S. Srivastava, P. Singh, A.K. Jagannatham, A. Karandikar, L. Hanzo, Bayesian learning-based doubly-selective sparse channel estimation for millimeter wave hybrid MIMO-FBMC-OQAM systems. IEEE Trans. Commun. **69**, 529–543 (2021). https://doi.org/10.1109/TCOMM.2020.3029568
19. Y. Xu, Z. Feng, J. Zou, D. Kong, Y. Xin, T. Jiang, An imaginary interference-free method for MIMO precoding in FBMC/OQAM systems. IEEE Trans. Broadcasting (2021). https://doi.org/10.1109/TBC.2021.3051528
20. Z. Yang, A. Zhang, A. Sudjianto, Enhancing explainability of neural networks through architecture constraints. IEEE Trans. Neural Netw. Learn. Syst. (2020). https://doi.org/10.1109/TNNLS.2020.3007259
21. X. You, C.X. Wang, J. Huang et al., Towards 6G wireless communication networks: vision, enabling technologies, and new paradigm shifts. Sci. China Inf. Sci. **64**, 110301 (2021). https://doi.org/10.1007/s11432-020-2955-6
22. R. Zakaria, D. Ruyet, On interference cancellation in Alamouti coding scheme for filter bank based multicarrier systems, in *Proceedings of the Tenth International Symposium on Wireless Communication Systems* (2013), pp. 1–5

Chapter 3
Deep Learning-Based Non-orthogonal Multiple Access for 5G and Beyond Networks

Mariyam Ouaissa, Mariya Ouaissa, Zakaria Boulouard, and Sarah El Himer

3.1 Introduction

Advances in mobile communication systems have seen remarkable development in relation to the number of users as well as the data traffic. Thanks to new generations of communication networks, the architecture has evolved and new techniques are proposed to improve the quality of services and the performance of networks for the implementation of new services [1]. The switch to 5G is seen as the convergence of Internet services with mobile network standards, offering what is known as mobile Internet over heterogeneous networks with very high connectivity speeds [2].

The high-level performance targets for 5G were developed as part of International Mobile Telecommunications (IMT) 2020, the International Telecommunication Union (ITU) initiative to define the basis for 5G. These requirements are associated with three major use cases, namely, Enhanced Mobile Broadband (eMBB), Massive Machine-Type Communications (MTC), and Ultra-Reliable Low-Latency Communication (URLLC) [3].

Unlike the previous standards such as 3G and 4G, which mainly aim to improve the speeds offered in mobility, 5G will be a transversal standard, which will aim to address a wide variety of connectivity issues [4]. To meet these requirements,

M. Ouaissa (✉) · M. Ouaissa
Moulay Ismail University, Meknes, Morocco
e-mail: mariyam.ouaissa@edu.umi.ac.ma; mariya.ouaissa@edu.umi.ac.ma

Z. Boulouard
Faculty of Sciences and Techniques Mohammedia, Hassan II University, Casablanca, Morocco
e-mail: zakaria.boulouard@fstm.ac.ma

S. El Himer
Faculty of Sciences and Techniques Fez, Sidi Mohammed Ben Abdallah University, Fes, Morocco
e-mail: sarah.elhimer@usmba.ac.ma

M. Ouaissa et al. (eds.), *Computational Intelligence in Recent Communication Networks*, EAI/Springer Innovations in Communication and Computing,
https://doi.org/10.1007/978-3-030-77185-0_3

improvements will first be made at the access network level, at the physical layer. To improve the spectral efficiency of 5G compared to 4G, the Non-Orthogonal Multiple Access (NOMA) [5] methods, that is, several users can use the same frequencies at the same time, are being considered. Discrimination between several users can be done by assigning different codes to each user. In the NOMA technique, all users can use the same spectrum and resources, thanks to the superposition coding (SC) mechanism at the transmitter and the Successive Interference Cancellation (SIC) at the receiver [6].

Channel State Information (CSI) is essential at the level of receivers to apply the SIC. Obtaining an ideal CSI at the transmitter level to design an efficient power allocation scheme is considered a delicate task. Deep learning techniques (DL) are proven to be an optimal solution for problems at the NOMA method level. In this work, we present a study of DL approaches and its advantages to improve the performance of NOMA systems. We discuss the principle of NOMA in the next section. Section 3.3 describes the main and most widely used deep learning techniques. In sect. 3.4, we specifically discussed the application of DL in NOMA. Finally, we conclude by conclusion.

3.2 System Model

Until now, successive generations of mobile networks have made it possible to increase the speed offered to users. However, the exponential growth in the number of connected objects is drawing more and more attention to their inability to meet the specific requirements of Machine-to-Machine (M2M) communications, also known as Machine-Type Communications. The traffic model of this type of communications, consisting of a large number of objects sporadically transmitting small packets, is the opposite of that considered for the design of mobile networks, namely, a small number of connections requiring a high bandwidth. As the objects must reach battery life of the order of several years, they also have very strong constraints in terms of energy consumption. This observation motivated the design of a fifth generation of cellular networks, having the will to support, among other things, this new paradigm, as well as a high degree of reliability and availability for Ultra-Reliable Communications (URC) [7].

A New Radio interface must therefore be developed, comprising innovative Physique (PHY) and Medium Access Control (MAC) layers components capable of addressing all 5G use cases. In Long-Term Evolution (LTE) networks, the design of which was driven by increasing demand for throughput, implementing sophisticated PHY and MAC layers providing a reliable and secure connection at the cost of significant but reasonable control traffic relative to the payload of a network human communication. For an object that would only communicate a few bytes, the relationship is reversed because the signaling then greatly exceeds the size of the data to be transmitted. The New Radio (NR) interface will have to implement new access mechanisms offering minimal overhead and reduced complexity on the object side. This can be expected to lead to increased complexity on the infrastructure side. The success of the future 5G radio interface for massive MTC support will essentially lie in the channel access strategy, namely, access with or without resource allocation, orthogonal or not, coding, etc. [8].

3.2.1 Principle

Conventional multiple access schemes consist of orthogonally separating users in at least one dimension (time, frequency, code, space) so that they do not interfere with each other. The number of possible users then depends on the quantity of available resources and the granularity of the scheduling. NOMA has recently been proposed as a transmission technique for the fifth generation of mobile communications systems and promises to dramatically increase spectral and power efficiencies over the current 4G standard, which uses Orthogonal Frequency Division Multiplexing (OFDM) [9]. In NOMA, multiple users are power multiplexed in the transmitter, over the OFDM layer, and signal separation takes place in the receiver using interference cancellation techniques SIC. The NOMA multiple access technique can perform at the uplink as well as the downlink [10]. Figure 3.1 illustrates the architecture of NOMA system.

The spectrum sharing for OFDMA and NOMA access techniques for two users is shown in Fig. 3.2.

This limit can be exceeded by waiving inter-user orthogonality. This then allows multiple users to use the same resources, therefore introducing "controlled" interference between the users. The separation of the different users then calls for sophisticated signal processing techniques exploiting a differentiator linked to the

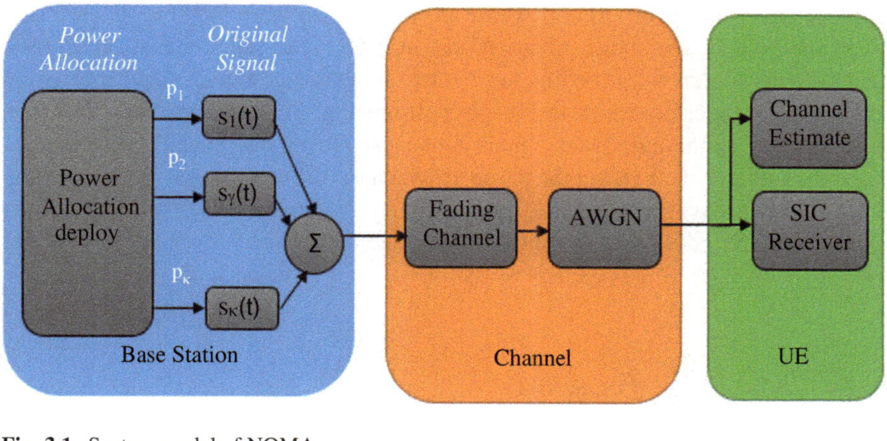

Fig. 3.1 System model of NOMA

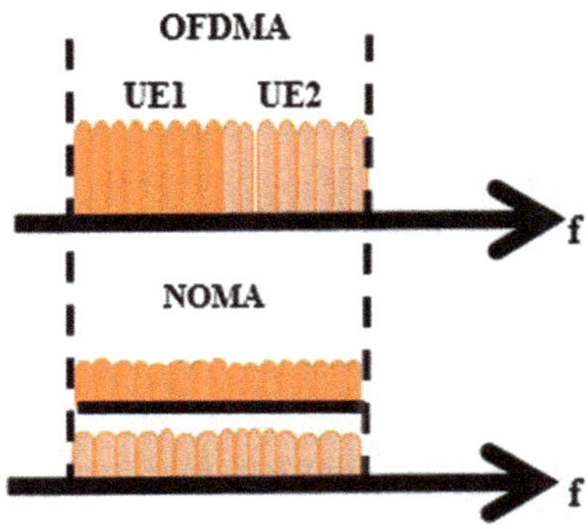

Fig. 3.2 Spectrum sharing for techniques OFDMA and NOMA (the case of two users)

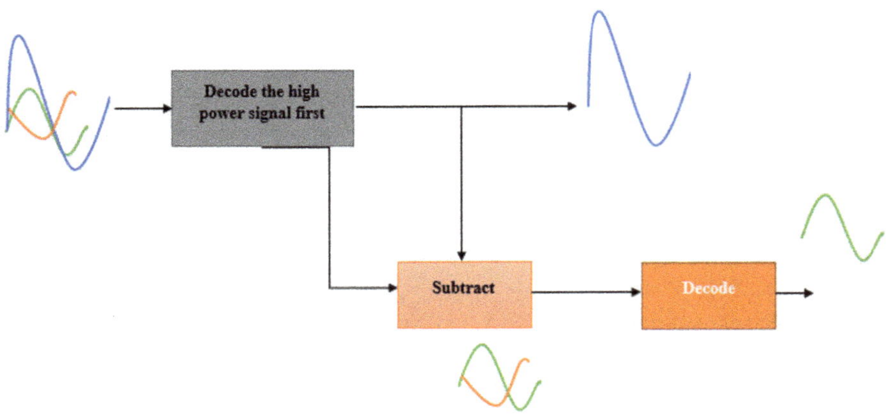

Fig. 3.3 Successive Interference Cancellation (SIC) scheme

radio channel. This is the case of a Successive Interference Cancellation receiver, capable of separating different power levels in power domain NOMA.

In the NOMA technique, all users can use the same spectrum and resources, thanks to the superposition coding mechanism at the transmitter and the cancellation of successive interference at the receiver. From Fig. 3.3, we can distinguish this principle. All of the individual signals are superimposed into a single waveform at the transmitter level, dividing the power between the different waves with precision. At the SIC receiver, the received signal includes all signals sent by the transmitter, and then the SIC decodes the signals one by one until it receives the desired signal [11].

3.2.2 *NOMA for Downlink*

The base station (BS) overlays user waveforms for a downlink NOMA system. Each User Equipment (UE) detects their signals using the SIC. According to Fig. 3.4, the

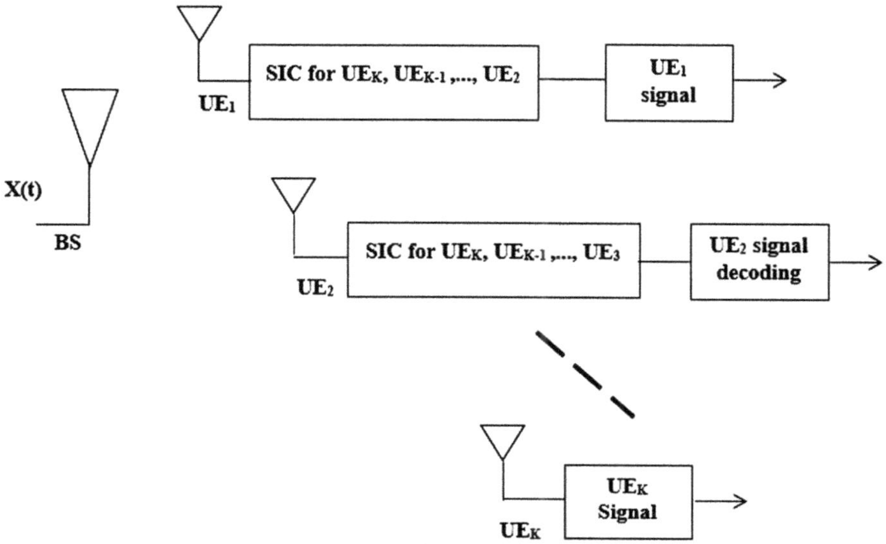

Fig. 3.4 NOMA downlink system

network is composed of a BS and a number K of users with SIC receivers where user 1 (UE_1) is the closest to the base station while user K (UE_k) is farthest. In the downlink, a base station faces the challenge of allocating power for different information waves, which poses a problem at the SIC level. In the network, the higher power is assigned to UE farthest from BS, while the lower power is assigned to the closest. First, each user decodes the strongest signal and then removes the decoded signal from the signal received, and this operation shall be repeated until the SIC receiver obtains its original signal [12].

The signal transmitted by the BS can be written as

$$x(t) = \sum_{k=1}^{k} \sqrt{\alpha_k P_T x_k(t)} \tag{3.1}$$

where x_X (t) is the individual information conveying the waveform, α_X is the UE_X power allocation coefficient, and P_T is the BS' total power. Each UE_X is then assigned $P_X = \alpha_X \times P_T$.

Power is allocated based on the distance of the UEs from the BS. UE_1 is closest to BS, so it is assigned the lowest power, while UE_K is the farthest, so its power is the highest. The signal received at the UE_X is

$$\mathrm{y}_k\ (\mathrm{t}) = \mathrm{x}\ (\mathrm{t})\ \mathrm{g}_k + \mathrm{w}_k\ (\mathrm{t}) \tag{3.2}$$

where g_X is the channel attenuation factor for the link between BS and UE_X, and w_X is the additive gaussian white noise to UE_X with zero mean and density N_0 (W/Hz).

First, take the most distant user. The signal that you first decode is its own signal because compared with the others, it gets the most power. The signals will be re-committed to other users as interference. Consequently, UE_X's SNR ratio is expressed

$$\mathrm{SNR}_k = \frac{P_k {g_k}^2}{N_0 W + \sum_{i=1}^{k-1} P_i {g_k}^2} \tag{3.3}$$

where W is the bandwidth of the transmission. The last signal that they decode will be a signal for the nearest UE_1. If cancelled perfectly, the SNR for UE_1 can be written as

$$\mathrm{SNR}_1 = \frac{P_1 {g_1}^2}{N_0 W} \tag{3.4}$$

In general, for UE_X, the SNR becomes

$$\mathrm{SNR}_k = \frac{P_k {g_k}^2}{N_0 W + \sum_{i=1}^{k-1} P_i {g_k}^2} \tag{3.5}$$

When NOMA is used, the rate (bps) for each UE can be written as

$$\mathrm{R}_k^{\mathrm{NOMA}} = \mathrm{W} \log_2 \left(1 + \frac{P_k {g_k}^2}{N + \sum_{i=1}^{k-1} P_i {g_k}^2} \right) \tag{3.6}$$

On the other hand, users in OFDMA are assigned to a group of subcarriers to receive the information. The throughput of each UE when the total bandwidth and power are shared between the UEs is expressed as

$$\mathrm{R}_k^{\mathrm{OFDMA}} = W_k \log_2 \left(1 + \frac{P_k |g_k|^2}{N_k} \right) \tag{3.7}$$

$$\text{where } W_k = \frac{W}{K} \text{and } N_k = N_o W_k$$

We also define the equity index which indicates how much the system capacity is shared between the UEs, that is, when F is close to 1, the capacity of each UE is close to one of the other.

$$F = \frac{\left(\sum R_k\right)^2}{k \sum {R_k}^2} \tag{3.8}$$

3.2.3 *NOMA for Uplink*

The uplink NOMA is somewhat different from its downlink implementation. A network multiplexing K UE in the uplink using NOMA can be displayed in Fig. 3.5. This time, the BS uses the SIC to differentiate user signals [13].

In the uplink, the signal received by the BS which includes all user signals is written as follows:

$$\mathrm{y}\,(\mathrm{t}) = \sum_{k=1}^{k} \mathrm{x}_k\,(\mathrm{t})\,\mathrm{g}_k + \mathrm{w}\,(\mathrm{t}) \tag{3.9}$$

where g_X is the channel attenuation gain for the link between BS and UE_X, x_X (t) is the information waveform for X^{th} UE, and $w(t)$ is the additive gaussian white noise at BS with zero mean and density N_0 (W/Hz).

With the uplink, UEs can further optimize their transmitting power by location, as in the downlink. However, we assume here that cellular coverage of customers is well distributed and that the power level of different users is already well separated. From a practical point of view, this assumption is more natural because power optimization requires a link that is difficult to implement between all the UEs.

The BS implements the SIC on the recipient. The first signal it decodes is the nearest user's signal, which always interferes with others. The SNR can be written

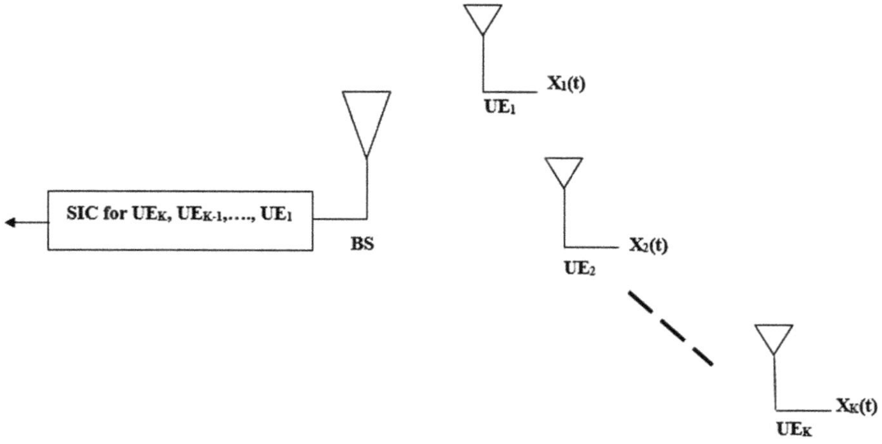

Fig. 3.5 NOMA uplink system

as follows for the signal for UE_1:

$$\mathrm{SNR}_1 = \frac{P{g_1}^2}{N + \sum_2^k P{g_i}^2} \tag{3.10}$$

where P is the power of the transmission of UEs and $N = N_0 W$.

The most remote user's UE_K is the final signal decoded by BS. If cancelled perfectly, the SNR for UE_K may be written as

$$\mathrm{SNR}_k = \frac{P{g_k}^2}{N} \tag{3.11}$$

Generally, for the Xth UE, the SNR becomes

$$\mathrm{SNR}_k = 1 + \frac{P{g_k}^2}{N + \sum_{i=K+1}^{k} P{g_i}^2} \tag{3.12}$$

The throughput for each UE can be expressed by

$$\mathrm{R}_k^{\mathrm{NOMA}} = \mathrm{W}\log_2\left(1 + \frac{P{g_k}^2}{N + \sum_{i=k+1}^{k} P{g_i}^2}\right) \tag{3.13}$$

On the other hand, in OFDMA, UEs are supported by orthogonal information. If the total bandwidth and power are shared equally between UEs, OFDMA generates

$$\mathrm{R}_k^{\mathrm{OFDMA}} = W_k \log_2\left(1 + \frac{P_k {g_k}^2}{N_k}\right) \tag{3.14}$$

where $W_k = \frac{W}{k}$ and $N_X = N_O W_X$.

To eliminate interference between users, two types of SIC receivers exist:

- Symbol-Level SIC (SLSIC) for detection without decoding of symbols that interfere with modulation.
- Codeword-Level SIC (CWSIC) for the detection and decoding of interfering data before its cancellation.

The CWIC recipient performs better than the SLIC recipient but is more complex.

3.2.4 Imperfection in NOMA

We have assumed that the SIC receiver will have a perfect cancelation. It is very hard to remove the decoded signal from a received signal without any errors from the real SIC receiver. Here we will introduce a cancelation error in the SIC receiver to the conception of NOMA. Since we already discussed in the uplink and downlink the concept of NOMA, we only look at the downlink to show the imperfection. Note that the SIC receiver is iteratively decoding the information signal to get the desired signal one by one. After decoding the signal, the original waveform should be regenerated to remove the signal from this receiver. Although this process can theoretically be completed without mistakes, some mistakes are expected to occur in practice [14].

In downlink, the SNR for the kth user with a rollback error is written as follows:

$$\mathrm{SNR}_k = 1 + \frac{P_k {g_k}^2}{N_0 W + \sum_{i=1}^{k-1} P_i {g_k}^2 + \in \sum_{i=k+1}^{k} P_i {g_k}^2} \tag{3.15}$$

where $\in$ is the error cancellation term which represents the remaining part of the canceled message signal. In the preceding case, the denominator's third term is not included because a perfect cancelation is assumed.

3.2.5 *Spectral and Energy Efficiency*

In this section, we analyze NOMA systems' spectral efficiency (SE) and energy efficiency (EE). In addition to the power consumed for the information waveform, we also include the network's static power consumption due to the power amplifiers. The sum of the energy of the information signal and the power-consuming circuits can be the total consumption of the transmitter mainly by power amplifiers [15]. In view of the downlink, the BS can then write the total power consumed as follows:

$$P_{totale} = P_T + P_{static} \tag{3.16}$$

where P_T is the total power of the signal mentioned above, and P_{static} is the power consumed by the circuit.

The EE is considered as the rate of the sum over the total power consumed of the base station:

$$EE = \frac{P_T}{P_{total}} = SE \frac{W}{P_{total}} \, (bits/joule) \tag{3.17}$$

where SE is the spectral efficiency (R_T/W) in terms of bps/Hz.

In Shannon's theory, the relation between EE and SE does not consider the circuit's power consumption, and it is thus monotonous that a higher SE always leads to a lower EE. When the circuit power is taken into account, the EE increases in the low region of SE and falls in the high area of SE. For a fixed P_{total}, the EE-SE relationship is linear with a positive R_T/P_{total} slope where an increase in SE results in an increase in EE simultaneously. NOMA offers more energy efficiency than OFDMA.

3.3 Overview of Deep Learning Models

In this section, the main deep learning techniques, namely, DNN, CNN, and RNN are presented [16].

3.3.1 Deep Neural Network

The term deep neural network (DNN) designates deep networks composed of several layers at least two are hidden and organized hierarchically (Fig. 3.6). Hierarchical organization allows information to be shared and reused. Along the hierarchy, you can select specific characteristics and eliminate unnecessary details to maximize invariance. The simplest structure that we know is that of feedback networks formed by back propagation algorithms. The updating of the weights during the learning phase is carried out using the SGD method, that is, to say in the formulas

$$w_{ij}(t+1) = w_{ij}(t) + \eta \frac{\partial C}{\partial w_{ij}} \tag{3.18}$$

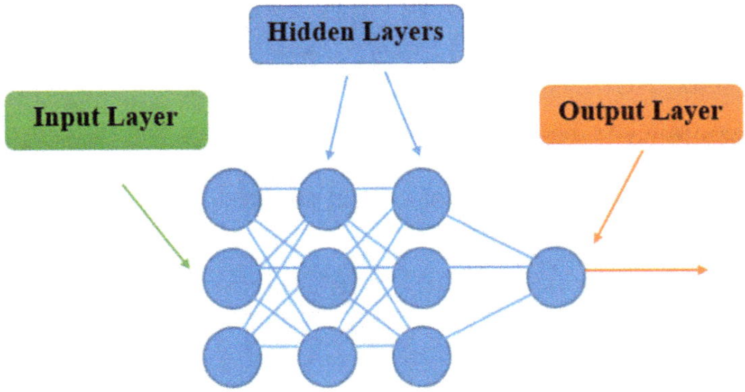

Fig. 3.6 Deep neural network (DNN) model

3.3.2 Convolutional Neural Network

Convolutional Neural Networks (CNNs) were introduced by Lecun et al. [17]. The peculiarity of CNNs is the use of the convolution operation in the first intermediate layers of the neural network. Originally, this operation was used as a filter in the area of image or sound to highlight patterns or reduce some type of noise. In CNNs, the model itself learns the filters of the different convolutions in order to highlight the patterns of the input data that are used in subsequent layers (Fig. 3.7). A classic CNN is generally made up of four types of layers:

- The convolutional layers, which contain several convolution operations applied to the same input.
- The layers of pooling operations.
- The activation layers.
- The layers all connected.

3.3.3 Recurrent Neural Network

While CNNs are primarily used to bring out spatially close relationships such as relationships between nearby pixels in an image, Recurrent Neural Networks (RNNs) have been developed to keep temporal context for each input event. They have been particularly used for the analysis of time series, audio data, or text where context is important in order to analyze each new entry. The idea is to keep information over time inside the layers of neurons to give context to the data being analyzed. The output of the RNN at time t will depend not only on the input at time t but also on the state of the RNN calculated at time t − 1. In its simplest version, a layer of an RNN can be described as a fully connected layer l which takes as input the previous layer l − 1 at time t concatenated at the output of itself, that is, layer l at instant t – 1.

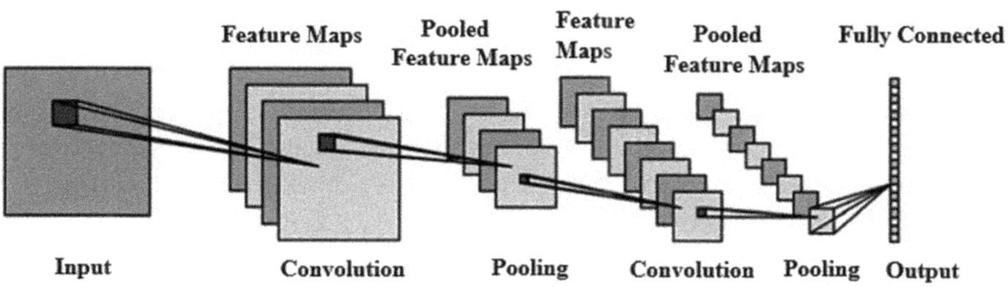

Fig. 3.7 Convolutional Neural Network (CNN) model

3.4 Deep Learning-Based NOMA

In wireless communications, the development of DL algorithms has brought significant performance benefits. Data from wireless systems, such as CSI, is increasingly heterogeneous and exhibits complex correlations. Efficient training of this data can show better performance [18].

In NOMA, the use of DL has been the subject of extensive research in recent years. Different types of learning algorithms are used in different aspects and demonstrate better performance compared to traditional schemes. In general, there are three types of learning methods. The first is called supervised learning where after presenting the desired data and results to computers, they will have the ability to make predictions for new input data. The second is unsupervised learning where the computer is given only data and where it has to find a structure with meaning on its own without the intervention of outside supervision. It mainly depends on clustering. The learning process takes place by observation and not by example. The third is reinforcement learning where the machine behaves like an agent learning from its environment in an interactive way until it discovers the behaviors that produce rewards [19].

DL is used in various applications in NOMA. For complicated data processing and for perfect CSI, the DL can be extremely useful. In NOMA, two types of training, such as online or offline training, can be identified. For training and recognition, the hidden layer is used. Hidden layers are particularly equipped with multiple neurons. Otherwise, one of the layers is named as noise layer, where the processed signal can be corrupted using additive-white gaussian noise. By extensive training of input data using existing channel models, the CSI can be automatically acquired [20]. A lot of training is carried out on arbitrary sequences of input signals by CSI from several simulation environment. This type of workout is called offline workouts. In addition, in order to accurately predict the channel environment, output signals are fed into the training data. The CSI can be transmitted by pilot signals in real time, and by this information, the input signals may be trained. Online learning is known as this type of learning. Figure 3.8 presents an online and offline training block diagram in the NOMA system for the self-detection of CSI. For the evaluation of the channel due to adverse weather, DL can also be used.

In addition, DL for signal decoding can be used with NOMA. There are several drawbacks to the conventional SIC method. Especially, when the number of users is increased, the SIC method is very difficult to decode the data perfectly. The propagation error also affects the SIC method. The deep neural network can retrieve a discrete sequence using signal classification from a deteriorated signal. The SIC method can also be optimized with DL.

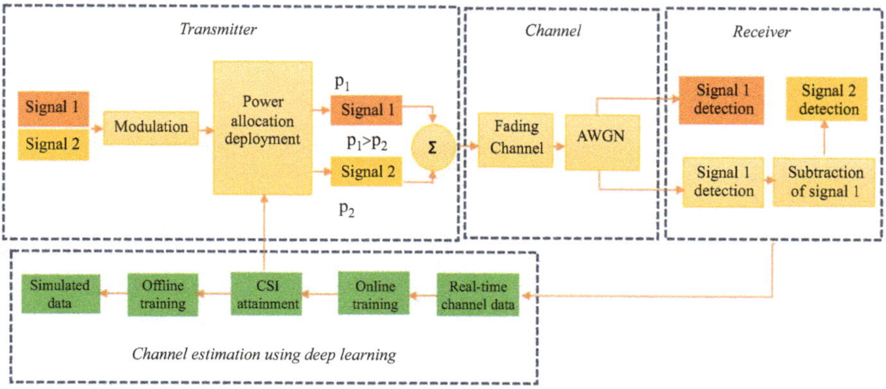

Fig. 3.8 DL-based NOMA technique using online and offline mechanisms

3.5 Conclusion

5G is the new generation of mobile telephony standards, this development will be spent on each physical object in the world, and it will be connected to the Internet under the new "IoT" concept. 5G is presented as the set of technologies whose goal is in particular to improve spectral efficiency by delivering more content at a higher rate. NOMA is recognized for its high spectral efficiency and low-latency features in 5G and beyond communication systems. Deep learning approaches can significantly improve its performance. The particular roles of the DL methods are briefly discussed in various NOMA applications in this study. It explains the way in which DL techniques improve NOMA performance.

References

1. M. Ouaissa, M. Ouaissa, A. Rhattoy, An efficient and secure authentication and key agreement protocol of LTE mobile network for an IoT system. Int. J. Intell. Eng. Syst. **12**(4), 212–222 (2019)
2. N. Panwar, S. Sharma, A.K. Singh, A survey on 5G: the next generation of mobile communication. Phys. Commun. **18**, 64–84 (2016)
3. M. Ouaissa, M Ouaissa, An improved privacy authentication protocol for 5G mobile networks, in *2020 International Conference on Advances in Computing, Communication & Materials (ICACCM)* (IEEE, 2020), pp. 136–143
4. M. Ouaissa, M. Houmer, M. Ouaissa, An enhanced authentication protocol based group for vehicular communications over 5G networks, in *2020 3rd International Conference on Advanced Communication Technologies and Networking (CommNet)*, IEEE, (2020), pp. 1–8.
5. L. Dai, B. Wang, Y. Yuan, S. Han, I. Chih-Lin, Z. Wang, Non-orthogonal multiple access for 5G: solutions, challenges, opportunities, and future research trends. IEEE Commun. Mag. **53**(9), 74–81 (2015)

6. L. Dai, B. Wang, Z. Ding, Z. Wang, S. Chen, L. Hanzo, A survey of non-orthogonal multiple access for 5G. IEEE Commun. Surv. Tutorials **20**(3), 2294–2323 (2018)
7. M. Ouaissa, A. Rhattoy, M. Lahmer, New method to control congestion for machine to machine applications in long term evolution system. Int. J. Commun. Antenna Propag. **8**(4), 355–363 (2018)
8. S. Li, L. Da Xu, S. Zhao, 5G internet of things: a survey. J. Ind. Inf. Integr. **10**, 1–9 (2018)
9. M. Ouaissa, A. Rhattoy, New method based on priority of heterogeneous traffic for scheduling techniques in M2M communications over LTE networks. Int. J. Intell. Eng. Syst. **11**(6), 209–219 (2018)
10. H. Tabassum, M.S. Ali, E. Hossain, M.J. Hossain, D.I. Kim, Uplink vs. downlink NOMA in cellular networks: challenges and research directions, in *2017 IEEE 85th vehicular technology conference (VTC Spring)*, IEEE, (2017), pp. 1–7
11. H. Sun, B. Xie, R.Q. Hu, G. Wu, Non-orthogonal multiple access with SIC error propagation in downlink wireless MIMO networks, in *2016 IEEE 84th Vehicular Technology Conference (VTC-Fall)*, IEEE, (2016), pp. 1–5
12. K.S. Ali, M. Haenggi, H. ElSawy, A. Chaaban, M.S. Alouini, Downlink non-orthogonal multiple access (NOMA) in Poisson networks. IEEE Trans. Commun. **67**(2), 1613–1628 (2018)
13. N. Ye, H. Han, L. Zhao, A.H. Wang, Uplink nonorthogonal multiple access technologies toward 5G: a survey. Wirel. Commun. Mob. Comput. **2018** (2018)
14. D.T. Do, T.T.T. Nguyen, Impacts of imperfect SIC and imperfect hardware in performance analysis on AF non-orthogonal multiple access network. Telecommun. Syst. **72**(4), 579–593 (2019)
15. H. Zhang, F. Fang, J. Cheng, K. Long, W. Wang, V.C. Leung, Energy-efficient resource allocation in NOMA heterogeneous networks. IEEE Wirel. Commun. **25**(2), 48–53 (2018)
16. O.I. Abiodun, A. Jantan, A.E. Omolara, K.V. Dada, N.A. Mohamed, H. Arshad, State-of-the-art in artificial neural network applications: a survey. Heliyon **4**(11), e00938 (2018)
17. Y. LeCun, L. Bottou, Y. Bengio, P. Haffner, Gradient-based learning applied to document recognition. Proc. IEEE **86**(11), 2278–2324 (1998)
18. H. Fourati, R. Maaloul, L. Chaari, A survey of 5G network systems: challenges and machine learning approaches. Int. J. Mach. Learn. Cybern. **12**(2), 385–431 (2021)
19. G. Gui, H. Huang, Y. Song, H. Sari, Deep learning for an effective non-orthogonal multiple access scheme. IEEE Trans. Veh. Technol. **67**(9), 8440–8450 (2018)
20. F. Kara, Ö.F. Gemici, İ. Hökelek, H.A. Çirpan, Optimal power allocation for DL NOMA systems, in *2017 25th Signal Processing and Communications Applications Conference (SIU)*, IEEE, (2017), pp. 1–4

Chapter 4
Traffic Sign Detection: A Comparative Study Between CNN and RNN

Amal Bouti, Mohammed Adnane Mahraz, Jamal Riffi, and Hamid Tairi

4.1 Introduction

Traffic sign recognition systems ensure that the current speed limit and other traffic signs are displayed to the driver on an ongoing basis. Automatic recognition works through a link between images captured by a vehicle-mounted camera and road signs stored in the navigation system. In this way, even signs that are not explicitly visible, such as in a city, will be displayed to the driver. In general, these systems recognize traffic signs to inform the driver by displaying a symbol representing the recognized sign. A camera installed on the car scans the side of the road for these signs. Driver assistance functions should not replace driver judgment or safe and skilled driving practices. The driver is solely responsible for monitoring vehicle conditions and the environment and for complying with applicable laws.

Autonomous cars are the future of the automotive industry [1, 2, 3, 4]. The intention of the design of these technological marvels is to incorporate efficient, driverless, and impact-resistant vehicles as these vehicles being intended for use in urban areas, with a lot of traffic, must be able to detect and recognize road signs to avoid accidents.

To our knowledge, there has not been a systematic comparison of CNN and RNN on the detection or recognition of traffic signs. The aim of these following paragraphs is to review the recent literature relating to both convolutional and recurrent neural networks, and especially long short-term memory.

Several researchers [5, 6, 7, 8] have used CNN to solve the problem of recognizing road signs. Convolutional neural networks are the tool of choice for these types of problems. These are the flagship algorithms of deep learning, objects

A. Bouti (✉) · M. A. Mahraz · J. Riffi · H. Tairi
LISAC, Faculty of sciences Dhar Al Mahraz Fez, Fes, Morocco
e-mail: amal.bouti@usmba.ac.ma

M. Ouaissa et al. (eds.), *Computational Intelligence in Recent Communication Networks* , EAI/Springer Innovations in Communication and Computing,
https://doi.org/10.1007/978-3-030-77185-0_4

of intense research, whose richness can impress. Recent studies in deep learning show how good a convolutional neural network is for image classification, and there are several advanced models with classification accuracies greater than 99%.

Recurrent neural networks are best known for solving time series predictions, while convolutional neural networks are known for image processing, and they are a strong candidate for sequence modeling as well as for time series forecasting [9].

As explained in [10], the importance of naming recurrent neural networks is that they use sequential information and perform the same task for each element of a sequence, resulting in the output being dependent on previous calculations. Recurrent neural networks have a memory that allows them to capture information in very long sequences, but they don't go back much. Recurrent neural networks are very efficient in sequential modeling units.

Tae-Young et al. [11] used a combination of a convolutional neural network, a deep neural network, and short-term and long-term memory which was given to them by the C-LSTM method. This method can automatically extract the robust features of spatiotemporal information from raw data, while experiments have shown that their method can extract more complex features. CNN layer is for reducing frequency variation of spatial information, DNN layer is for mapping data into more separable space, and LSTM layer is suitable for modeling temporal information.

Convolutional neural networks and recurrent neural networks are the two main types of deep neural networks. They have the ability to handle various tasks of natural language processing; hence, DNNs have made a revolution in this field. CNNs are efficient in the part of extracting position invariant entities, and RNNs in the part of sequential modeling units [12].

Many advanced algorithms based on neural networks have been developed to solve this problem. In this project, two such algorithms are identified, namely, the convolutional neural network-based approach [13] and the recurrent-based approach [14], for the detection of traffic signs. Both algorithms are trained using traffic sign images and are tested on an input image to give an output, which is a prediction of what the input image is. However, rather than a simple prediction, the performance of the two methods is compared and analyzed in this project.

To do this, we have structured our Chapter in 3 sections: In the first section, we will present the different methods used in our system as well as their interests in the field of image classification. In the second section, we will do a small comparison between convolutional neural network and recurrent neural network, and we will describe our work. Finally, in the third section, we will show the different results obtained, and at the end, we end with a general conclusion.

4.2 Materials and Methods

4.2.1 Convolutional Neural Networks

Convolutional neural networks are a special type of feedback neural network that draws its inspiration from biological processes occurring in the optic lobe specifically in the brains of living organisms and is believed to be a solution to many computer vision problems. In artificial intelligence such as image and video processing, its structure has several designs. Le-Net 5 is the most popular and practical model and was introduced by Yann LeCun in 1998. It is necessary to familiarize ourselves with the basic concepts of the artificial neural network.

4.2.1.1 Multilayer Neural Networks

A neural network is made up of a group of neurons arranged in layers consisting mainly of an input layer, an output layer, and one or more hidden layers.

4.2.1.2 Feed-Forward Neural Network

It is one of the most important types of multilayer neural networks, so named because it adopts the principle of direct propagation, where the output of all neurons in layer m is the input of each neuron in the layer in front of him m + 1.

4.2.1.3 Learning and Training

The learning phase of the neural network is an essential and important phase that comes after its design phase, where it is trained on a set of examples to give the network correct results on all the examples we have trained on, but the training process does not stop At this limited objective, the privilege of the idea of neural networks rather passes through the process of generalization, that is, the capacity of the network to find the right outcome for new examples he has not been trained on before. The most common algorithm for training neural networks is back propagation. It is an iterative process that starts from the last layer, calculates the error obtained by the output by finding the difference between its value and the value of the required output, and then tries to reduce the value of that error using the stochastic gradient descent algorithm or other optimization algorithms that modify the values of the output layer weights in a way that reduces the error obtained. Then this process is repeated on the penultimate layer and so on until the first layer. Then all of the above can be repeated on all the layers to achieve the optimal weights that reach the minimum acceptable value of the network error.

4.2.1.4 Structure of a Convolutional Neural Network

The neurons of this type of network are arranged in layers of different types (Fig. 4.1).

4.2.1.5 Convolutional Layers

The naming of this type of layer comes from the mathematical process of folding or wrapping, which is a process applied to the functions f and g and produces the function o, which is a modified version of the two methods and gives the area of interference between them. Mathematical wrapping on the input elements (neurons of the previous layer or input pixels of the image) with a number of filters or wrapping of kernels and the previous wrapping process is repeated on each group of elements d, input with the size of a single filter, to produce one of our feature map or feature map values. Then we repeat all the previous operations for each filter to produce a map of the other features and so on. The values of these filters express the weights of the grid. We observe the following in these layers.

4.2.1.6 Shared Weights

Repeated application of the filter wrapping process on the whole of the income element will greatly improve the performance as it will lead to the presence of a large number of connections, but with common weights between them, and that is one of the most important characteristics of this type of networks because it increases the efficiency of the learning and makes the network eligible to obtain better results in generalization. For example, each resulting neuron in the characteristic map will be connected by connections to the number of filter elements, but the weights of those connections will be the same as the weights of the neighboring neuron in the same feature map, and this process will also secure recognition of the features of the image by looking at its location in the image, base image because it connects the adjacent dots to each other and also allows the representation of regions.

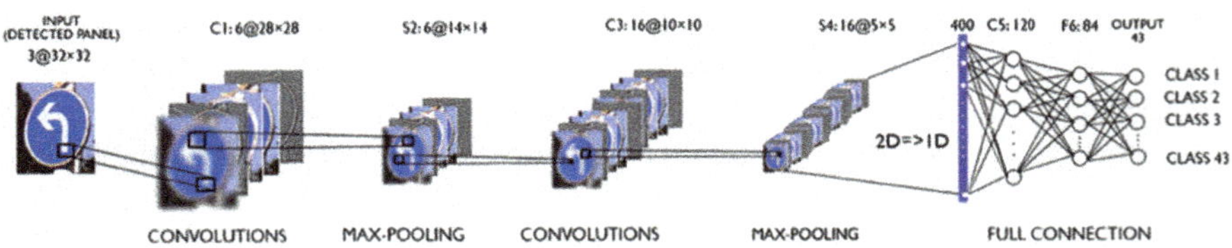

Fig. 4.1 Example of CNN architecture [15]

4.2.1.7 Multiple Filters

Applying multiple filters to the folding process with the same input will identify different groups of features in the image since each of them will produce a unique feature map, but their number should be placed so as not to not neglect the complexity of the current calculations as it depends on the number of examples available for training and the complexity, assignment, contrast, and dimensions of the input image while the size of a single candidate, that is, its dimensions, depends mainly on the data from the set of training examples.

4.2.1.8 Subsampling Layers

These layers are optional in the network design, that is, their presence is not required, and if they exist, they will come after each of the convolutional layers and aim to reduce the number of samples or neurons, because you will shorten each neuron input group by a neuron-specific size, and this size is determined in the design of the network, and its optimum value is 2×2 because its enlargement can result in loss of information, and the reduction is done in several ways, including:

Max pooling: takes the highest value among them.
Average pooling: takes the average of all values.

4.2.1.9 Fully Connected Layers

After several layers of the two preceding types, these layers come to connect all the neurons of the preceding layer (whatever their type) and to make an entry for each neuron in it as in the regular neural networks, it is not necessary be it of a number, but often there are two consecutive layers as the last layers of the network because it cannot have come before a convolutional layer.

4.2.1.10 Correction Layers (ReLU)

The main goal is to interpose a layer, which performs an activation function on the output signals, between the layers to improve processing efficiency. We have in particular:

The ReLU correction: $f(x) = \max(0, x)$.
Correction by hyperbolic tangent $f(x) = \tanh(x)$.
Correction by the saturating hyperbolic tangent: $f(x) = |\tanh(x)|$.
Correction by the sigmoid function $f(x) = (1 + $e-$x$)-1.

Often, the ReLU correction is preferable as it results in neural network formation several times faster [16], without making a significant difference to generalization precision. Its variants are summarized in the following table (Table 4.1).

4.2.1.11 Loss Layer (LOSS)

This layer is usually the last layer of the network. It gives the difference between the expected and real signal by describing the slip of the network. There are several loss functions from which each function is adapted to a spot. The "Softmax" loss is used to predict a single class among K mutually exclusive classes. The sigmoid cross entropy loss is used to predict K independent probability values in [0, 1]. Euclidean loss is used to regress to real values in $[-\infty, \infty]$.

4.2.2 *Recurrent Neural Networks*

Recurrent neural networks are a type of neural networks that allow past predictions to be used as inputs through hidden states. They are of the following form (Fig. 4.2):

At time t (Fig. 4.3), the activation $a^{<t>}$ and the output $y^{<t>}$ are the following form:

$$a^{<t>} = g_1\left(W_{aa}a^{<t-1>} + W_{ax}x^{<t>} + b_a\right)\ et\ y^{<t>} = g_2\left(W_{ya}a^{<t>} + b_y\right)$$

where W_{ax}, W_{aa}, W_{ya}, b_a, b_y are time-independent coefficients and where g_1 and g_2 are activation functions.

Table 4.1 The variants of the rectified linear unit layer

Times	Leaky ReLU	ELU
$g(z) = \max(0,z)$	$g(z) = \max(\epsilon z, z)$ with $\epsilon \ll 1$	$g(z) = \max(\alpha\,(ez - 1),z)$ With $\alpha \ll 1$
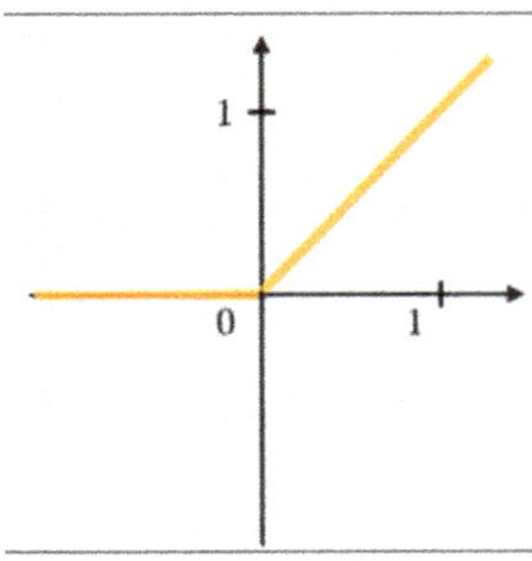	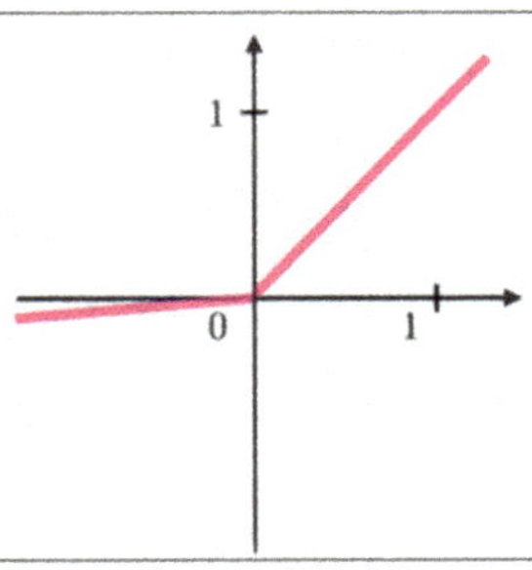	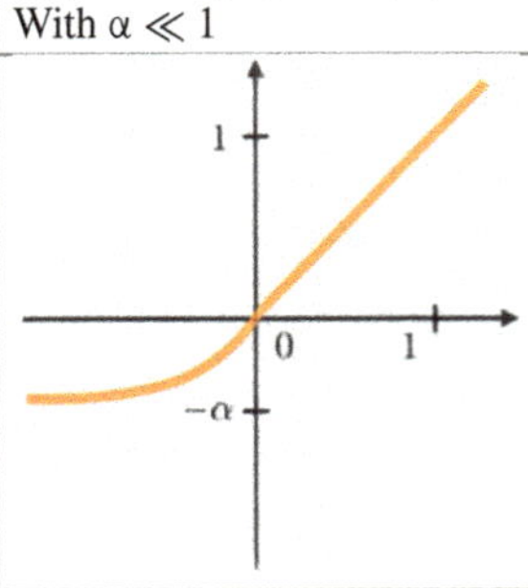
Nonlinear complexities interpretable from a biological point of view	Responds to ReLU dying problem	Derivable everywhere

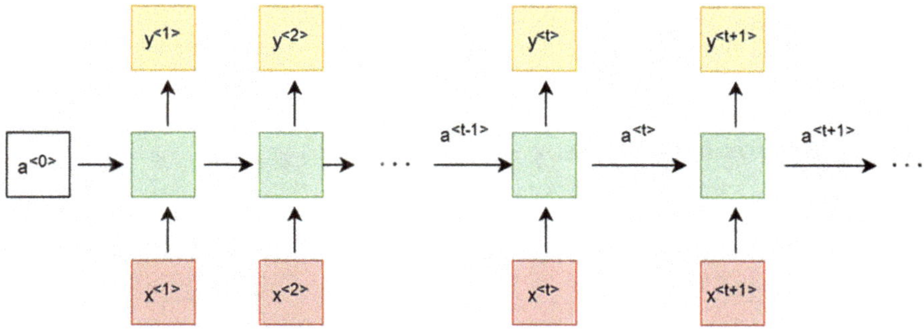

Fig. 4.2 Recurrent neural network

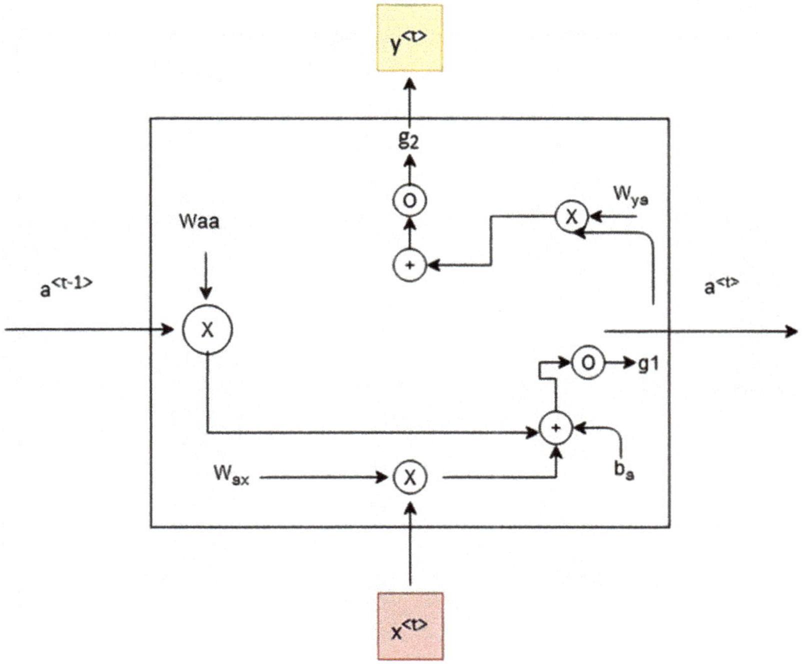

Fig. 4.3 The recurrent neural network at time t

The advantages of traditional RNN architectures are:

- The possibility of taking into account entries of any size.
- The size of the model does not increase with the size of the entry.
- The calculations take into account previous information.
- The coefficients are independent of time.

Among the disadvantages of traditional RNN architectures are:

- The calculation time is long.

- Difficulty accessing information from the distant past.
- Impossibility of taking into account future information in a given state.

4.2.2.1 Applications of RNNs

RNN models are mainly used in the fields of automatic natural language processing and speech recognition. The following table details the main applications to remember (Table 4.2):

4.2.2.2 Loss Function

In the context of recurrent neural networks, the loss function L takes into account the loss at each time T as follows:

$$\mathcal{L}\left(\hat{y}, y\right) = \sum_{t=1}^{T_y} \mathcal{L}\left(\hat{y}^{<t>}, y^{<t>}\right)$$

Table 4.2 The main applications of RNNs

RNN type	Illustration	Example
One to one $T_x = T_y = 1$	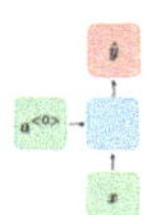	Traditional neural network
One to many $T_x = 1, T_y > 1$	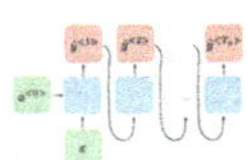	Music generation
Many to one $T_x > 1, T_y = 1$	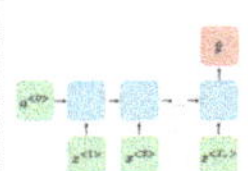	Sentiment classification
Many to many $T_x = T_y$	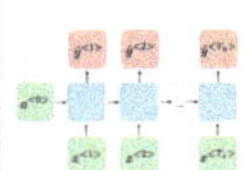	Entity recognition
Many to many $T_x \neq T_y$	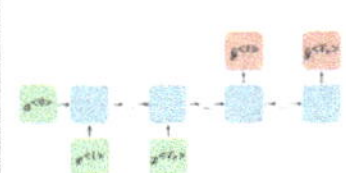	Machine translation

4.2.2.3 Temporal Back Propagation

The back propagation step is applied in the time dimension. At time T, the derivative of loss L with respect to the matrix of coefficients W is given by.

$$\frac{\partial \mathcal{L}^{(T)}}{\partial W} = \sum_{t=1}^{T} \left. \frac{\partial \mathcal{L}^{(T)}}{\partial W} \right|_{(t)}$$

4.2.2.4 Activation Functions

The activation functions most used in RNNs are described below (Table 4.3):

4.2.2.5 Long Short-Term Memory

A Long Short-Term Memory Network (LSTM) was proposed in 1997 by Sepp Hochreiter et al. [17]. It is the architecture of neural networks the most used which makes it possible to solve the problem of the disappearance of the gradient, and each calculation cell of this architecture is linked to a hidden state h and also to a state c of the cell that plays the role of memory. LSTMs have units that act as building units for the layers of a recurrent neuron network. LSTMs have memory very similar to computer memory which allows them to store, write, read, and delete their information and to remember their entries over a long period of time. LSTMs have three gates: input, forget, and output.

You can see an illustration of an RNN with its three gates below (Fig. 4.4).

4.3 Proposed System

4.3.1 Comparison Between CNN and RNN

Now we will focus on the difference between convolutional neural networks and recurrent neural networks where in this table we compare the two neural networks

Table 4.3 Activation functions

Sigmoid	Tanh	ReLU
$g(z)=\frac{1}{1+e^{-z}}$	$g(z)=\frac{e^{z}-e^{-z}}{e^{z}+e^{-z}}$	$g(z)=\max(0,z)$
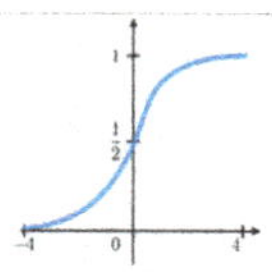	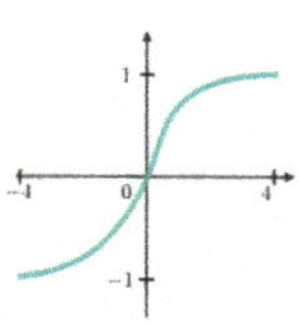	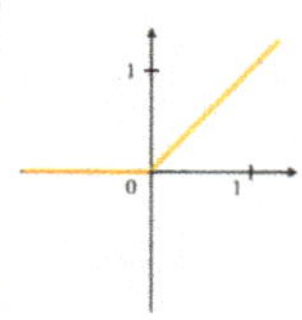

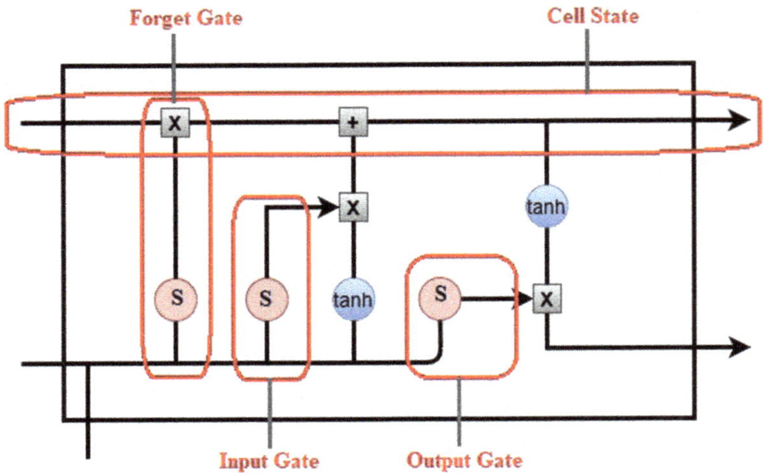

Fig. 4.4 An illustration of LSTM

Table 4.4 A comparison between CNN and RNN

	Convolutional neural network	Recurrent neural network
Architecture	Feed-forward neural networks using filters and pooling	Recurring network that returns results in the network
Input/output	The size of the input and the resulting output are fixed (i.e., receives images of fixed size and outputs them in the appropriate category with the confidence level of its prediction)	The size of the input and the resulting output may vary (i.e., receives different text and output translations; the resulting sentences may have more or less words)
Ideal usage scenario	Spatial data (such as images)	Temporal data/sequential (such as text or video)
Use cases	Image recognition and classification, face detection, medical analysis, drug discovery, and image analysis	Text translations, natural language processing, language translation, entity extraction, conversational intelligence, sentiment analysis, and speech analysis

with respect to architecture, input, output, ideal usage scenario, and use cases (Table 4.4).

4.3.2 *Our System*

Since both neural networks must first be trained, a dataset is needed. Here, the dataset is taken from the German Recognition Benchmark database, which is available for open-source download from Kaggle [18]. The dataset contains 43 different classes of road signs, such as speed limits, stop signs, etc. There are a total of 39,209 images from the training dataset spread across the 43 traffic signs, which means approximately 906 images for each type of sign. There are 12,630 images in the test dataset. Each image has been resized to a size of 30×30 to minimize computational complexity as much as possible. For each image, the pixel values are scaled by dividing by 255 for better normalization. The training data is divided into a breakdown of 80% for the training set and 20% for the validation set. The following image (Fig. 4.5) shows the 43 different traffic signs. It can be noted that for a particular class of road signs, different types of images were collected and stored in the dataset.

As for the software to use, both neural networks, CNN and RNN-LSTM, can be implemented in python using TensorFlow and its subfunctions. We used Google Colab. We initially assign the two neural networks three hidden layers, the ReLU activation function, and a dropout rate of 0.25. The number of epochs is fixed at 10. A dataset containing images of various road signs is obtained. This dataset is provided to each neural network. The training is performed until both networks can recognize a given sign input to a reasonable extent. The main objective is to observe the behavior of each neural network when it is subject to change in three crucial network parameters, namely, the number of neurons in each layer, the type of activation function used, and the dropout percentage for each layer. Changes in precision, F1 score, and complexity for each neural network are observed when

Fig. 4.5 Samples from the German Traffic Sign Recognition Benchmark. Image from [18]

the aforementioned parameters are changed, and therefore, an observation on the performance of each neural network is made.

We first change the number of neurons in both networks, keeping the activation function and the dropout rate constant. Then we repeat the same, changing the dropout rate for both neural networks, keeping the other two parameters constant, and finally changing the activation function for both. The impact of these changes on accuracy, F1 score, and complexity are recorded. For each parameter change, the code is run four times on python for each neural network, and the average precision, F1 score, and complexity are calculated. The code is executed several times in order to verify the authenticity of the results obtained and to ensure that the results obtained are certain and do not change much.

4.4 Results Obtained

The following section shows the precision and loss curves for training and validation, for CNN and RNN, and provides two tables which contain the different configurations and results obtained after training on the two neural networks.

The curves presented in the figure (Fig. 4.6) concern the performance of the two methods when the network parameters have been set to the default values. In other words, the parameter configuration has been defined as follows: P1 in Table 4.5.

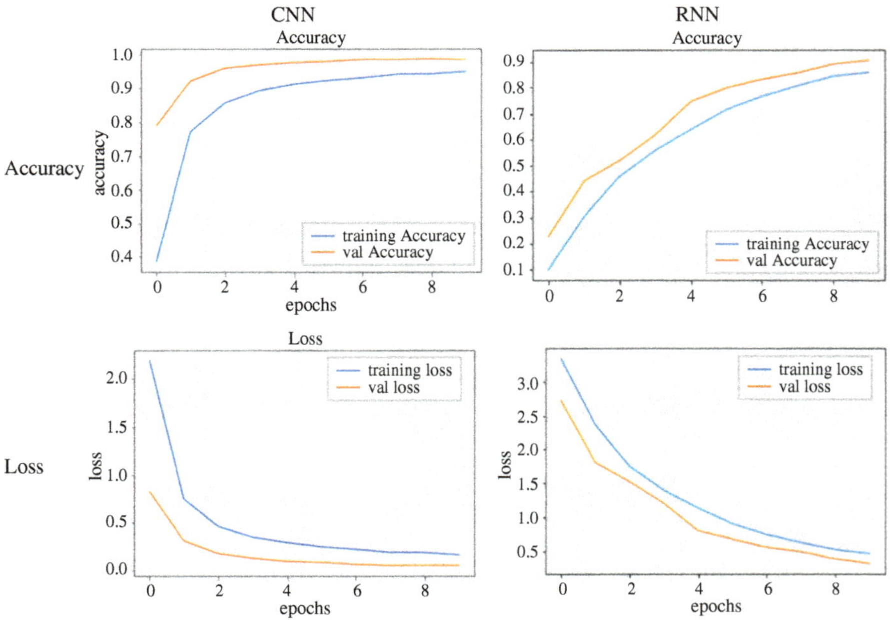

Fig. 4.6 Accuracy curves and loss for the training and validation for CNN and RNN

Table 4.5 Configuration of the parameters

	Neurons in H1	Neurons in H2	Neurons in H3	Dropout rate	Activation function
P1	32	64	64	0.25	ReLU
P2	128	128	64	0.25	ReLU
P3	32	64	64	0.15	ReLU
P4	32	32	64	0.4	ReLU
P5	32	64	64	0.25	Sigmoid
P6	32	64	64	0.25	TanH

Table 4.6 The results obtained for the accuracy, F1 score, and complexity for the different configurations of the parameters mentioned in the fifth table

	Measures	CNN	LSTM
P1	Accuracy	94.75%	74.53%
	F1 score	92.10%	63.75%
	Complexity	210.3 s	273.25 s
P2	Accuracy	92.4%	84.53%
	F1 score	90.24%	75.75%
	Complexity	864.5 s	573.25 s
P3	Accuracy	94.4%	78.75%
	F1 score	91.5%	70.75%
	Complexity	280.73 s	283.45 s
P4	Accuracy	90.4%	62.75%
	F1 score	88.5%	44.75%
	Complexity	264.3 s	285.7 s
P5	Accuracy	82.4%	24.45%
	F1 score	74.5%	9.5%
	Complexity	266.39 s	297.88 s
P6	**Accuracy**	**95.7%**	**85.4%**
	F1 score	**93.5%**	**79.5%**
	Complexity	**275.38 s**	**300**

From the second table, we can deduce that the change of the activation function to Tanh caused the RNN to offer its best performance, both in terms of precision and complexity scores. CNN's performance was improved by this operation. From there, some interesting observations can be deduced. From the results obtained for the two implementations, the general tendency is that CNN is more robust when it comes to modifying network parameters, that is, its performance changes only slightly. On the other hand, the RNN turns out to be more sensitive because there have been situations where the accuracy was extremely low, as well as relatively high with less complexity, for small changes in parameters, but in no case does this mean that RNN is not a good choice. It just means that one has to be extremely careful when designing an RNN for their application as small changes in the network parameters can cause huge differences in the desired outputs (Table 4.6).

4.5 Conclusion

Finally, it can be concluded that the fundamental step of detecting traffic signs was performed with two neural networks: A dataset containing the signs was fed to both networks and trained. The effect of changing network parameters on the performance of the two networks was recorded an inference about how each neural network behaves in relation to each other was made, which tells us how the two methods behave in relation to each other under different conditions and circumstances. In the future, now that we have noted the behavior of recurrent neural network versus convolutional neural network, we would like to expand our research by trying a combination of CNN and RNN on real-world video streams containing traffic signs and tuning neural network parameters in favor of less complexity and greater precision. Additionally, we aspire to observe more trends in the performance of the two neural networks, exploring more network parameters to modify.

References

1. J. Ondruš, E. Kolla, P. Vertaľ, Ž. Šarić, How do autonomous cars work? Transp. Res. Procedia **44**, 226–233 (2020)
2. A. Gupta, A. Anpalagan, L. Guan, A.S. Khwaja, Deep learning for object detection and scene perception in self-driving cars: survey, challenges, and open issues. Array **100057** (2021)
3. N.G.S.S. Srinath, A.Z. Joseph, S. Umamaheswaran, C.L. Priyanka, M. Nair, P. Sankaran, NITCAD-developing an object detection, classification and stereo vision dataset for autonomous navigation in Indian roads. Procedia Comput Sci **171**, 207–216 (2020)
4. E. Khatab, A. Onsy, M. Varley, Vulnerable objects detection for autonomous driving: a review. Integration (2021)
5. A. Vennelakanti, S. Shreya, R. Rajendran, D. Sarkar, D. Muddegowda, P. Hanagal, Traffic sign detection and recognition using a CNN ensemble, in *2019 IEEE international conference on consumer electronics (ICCE)* (2019), (pp. 1–4). IEEE
6. Z. Liu, J. Du, F. Tian, J. Wen, MR-CNN: a multi-scale region-based convolutional neural network for small traffic sign recognition. IEEE Access **7**, 57120–57128 (2019)
7. A. Bouti, Mahraz, M. A., Riffi, J., Tairi, H., Road sign recognition with Convolutional Neural Network, in *2018 International Conference on Intelligent Systems and Computer Vision (ISCV)* (2018), (pp. 1–7). IEEE
8. W.A. Haque, S. Arefin, A.S.M. Shihavuddin, M.A. Hasan, DeepThin: a novel lightweight CNN architecture for traffic sign recognition without GPU requirements. Expert Syst. Appl. **168**, 114481 (2021)
9. T. Nguyen, G. Nguyen, B.M. Nguyen, EO-CNN: an enhanced CNN model trained by equilibrium optimization for traffic transportation prediction. Procedia Comput. Sci. **176**, 800–809 (2020)
10. F. Zanetti. Convolutional networks for traffic sign classification. MS thesis. (2016)
11. T.Y. Kim, S.B. Cho, Web traffic anomaly detection using C-LSTM neural networks. Expert Syst. Appl. **106**, 66–76 (2018)
12. W. Yin, K. Kann, M. Yu, H. Schütze, Comparative study of CNN and RNN for natural language processing. arXiv preprint arXiv:1702.01923 (2017)
13. S. Albawi, T.A. Mohammed, S. Al-Zawi, Understanding of a convolutional neural network, in *2017 International Conference on Engineering and Technology (ICET)*, (2017), (pp. 1–6). Ieee

14. A.N. Michel, Recurrent neural networks: overview and perspectives, in *Proceedings of the 2003 International Symposium on Circuits and Systems, 2003. ISCAS'03*. (Vol. 3, pp. III–III). (2003), IEEE
15. A. Bouti, M.A. Mahraz, J. Riffi, H. Tairi, A robust system for road sign detection and classification using LeNet architecture based on convolutional neural network. Soft Comput., 1–13 (2019)
16. A. Krizhevsky, I. Sutskever, G.E. Hinton, Imagenet classification with deep convolutional neural networks. Adv. Neural Inf. Proces. Syst. **25**, 1097–1105 (2012)
17. J. Schmidhuber, S. Hochreiter, Long short-term memory. Neural Comput. **9**(8), 1735–1780 (1997)
18. https://www.kaggle.com/meowmeowmeowmeowmeow/gtsrb-german-traffic-sign

Chapter 5
Merging Attack-Defense Tree and Game Theory to Analyze Vehicular Ad Hoc Network Security

Meriem Houmer, Mariya Ouaissa, Mariyam Ouaissa, and Moulay Lahcen Hasnaoui

5.1 Introduction

Over recent years, automobiles have ceased to be considered a means of transport to become mobile data centers to increase drivers and passenger safety, also to prevent accidents. For this, a wide variety of mobile applications have been designed, which has created a new market, due to the need to transfer information from vehicles or occupants through the Internet or another type of network. The new generation of vehicles is able to make connections with their surroundings through what has become known as vehicular ad hoc networks.

Vehicular ad hoc networks (VANET) are a subclass of mobile ad hoc networks (MANET) in which vehicles can communicate with each other shaping vehicle to vehicle communication (V2V) or communicating with the roadside units (RSU) shaping vehicle to infrastructure communication (V2I) [1]. This network has been designed to provide several advantages, such as improve traffic efficiency, minimize traffic congestion, avoid accidents, and make easy access to news, information, and entertainment while driving [2]. However, VANET covers a broad variety of applications, which can be split between security services, traffic management, and user-oriented services. All of these applications require the exchange of messages such as emergency messages, traffic incidents warnings, and road conditions at real times and driving assistance information [3]. These exchanges demand real-

M. Houmer (✉) · M. L. Hasnaoui
Information and Communication Systems Engineering Research Team, Mathematical Modeling and Computer Science Laboratory, National Graduate School of Arts and Crafts, Moulay-Ismail University, Meknes, Morocco

M. Ouaissa · M. Ouaissa
Moulay Ismail University Meknes, Meknes, Morocco
e-mail: mariya.ouaissa@edu.umi.ac.ma; mariyam.ouaissa@edu.umi.ac.ma

M. Ouaissa et al. (eds.), *Computational Intelligence in Recent Communication Networks* , EAI/Springer Innovations in Communication and Computing,
https://doi.org/10.1007/978-3-030-77185-0_5

time communication data from all nodes in the network because the content of messages can influence on drivers' behavior. This, therefore, implies a risk of attack by malicious users who can alter or delete messages exchanged on the network [4]. Attackers can prohibit the routing of different packets by targeting network availability at the different layers of the protocol pile and then make the major aim of VANET ineffective. Unfortunately, the limitations created by high-node mobility (highly dynamic topology) and the decentralized aspect of vehicular ad hoc networks make its security more problematical than every other type of network.

The existing VANET security research is typically based on prevention strategies that lack a specified assessment to enable the administrator of the network to rank the most potential attacks and determine efficient defense techniques to success the vehicular system. Moreover, attack and defense mechanisms are directly related, which implies that the risk assessment should include both the attacker and the defender strategies. Actually, the attacker and the defender aim to maximize their benefits rather than automatically generate a blind attack or establish preventive measures.

Our paper aims to present a simplified process for protecting VANET availability. Contrary to the existing approaches, our research will not only explore how an attack can be avoided, but it includes the costs of each attack launched by the attacker as well as the cost of the defender's countermeasure, respectively, their gains. First of all, we introduce an approach called attack-defense tree to model the different possible attacks that can be launched by the attackers, also their corresponding countermeasure chosen by the defender. Then we utilize the mathematical modeling of strategic interaction among rational decision-makers, named the game theory [5], to analyze the efficiency of our attack-defense tree. Here, the application of game theory is to model and evaluate the various relationships between attacker and defender. Moreover, we introduce the Return on Attack (ROA) that represents the potential gain that the attacker can get from an attack led. We provide also the Return on Investment (ROI) that represents the defender's potential gain through adopting countermeasures. These two functions are considered as the key components of our attack-defense game since each player (attacker or defender) tries to maximize the utility of its strategy by maximizing the ROI or the ROA. Based on the findings, we are able to determine the appropriate security strategies to make the system safer.

The rest of the paper is organized as follows: The system model is described in Sect. 5.2. Section 5.3 presents the attack-defense tree fundamental. Section 5.4 carries out our attack-defense tree model for VANET availability. The economic factors (Return on Attack (ROA), Return on Investment (ROI)) are introduced in Sect. 5.5. Section 5.6 presents the basics of game theory as well as the analysis of our attack-defense game. Finally, we conclude this article in Sect. 5.7.

5.2 System Model

Vehicular ad hoc network represents an autonomous and self-organized wireless communication network, where vehicles act as mobile nodes, to provide communication between vehicles fitted with onboard unit (OBU), also between vehicles and the roadside units (RSU). We adopt the following assumptions [6]:

- Each vehicle in VANET is fitted with an OBU (onboard unit), allowing communication between vehicles and to communicate with the roadside units (RSU).
- A trusted third authority named the certificate authority (CA) is assumed to exist. It represents a transportation authority that is responsible for the security and privacy of the network. It is truthfully performing cryptographic operations, including key generation operations.
- The CA distributes a unique identity to each vehicle in the network and a set of pseudonyms. When the vehicle is registered at the CA, it becomes a legitimate entity. The validity of the node's identity is verified periodically.
- The pseudonym of each vehicle is changed after an interval of time in other to ensure privacy preservation. However, if a vehicle uses their entire pseudonym, the CA is charged to issue it a new set of pseudonyms.
- Vehicles possess sufficient energy and storage capacity to set up and operate an individual firewall or antivirus program to safeguard them against malware attacks and malicious software such as viruses that propagate through wireless networks.
- RSU is not trusted because it can be easily compromised as it is located along the road. Nevertheless, the RSU and the CA communicate across a secure fixed network.

5.3 Fundamental Attack-Defense Tree

The attack-defense tree is a methodology explaining how an attacker can act to lead an attack against a system and also the different protection mechanisms that the defender can use to defend it. In general, the attack-defense tree provides a global objective (attacker goal) that represents the tree's root node and many subgoals that represent the leaf nodes. It has two distinct nodes: the attack node and the defense node. These nodes are corresponding to the subgoals of the attacker and of the defender. Refinement and countermeasures represent two main features of the attack-defense [7]. Each node could have one child or more of the same kind that refines the global goal to subgoals. If a node has no child of the same type, it is referred to as an unrefined node, and therefore, it represents a basic action. However, each node could also have opposing children, which represents the countermeasures. Moreover, an attack node could have many children who optimize it and only one countermeasure child that defend it. In their turn, the countermeasure child may have many children that optimize the defense. Two types of optimization (operators) are usable: conjunctive refinement (AND) and disjunctive refinement (OR) [8, 9]. These operators are regarded as the fundamental gates for building the tree. To attain a goal of an AND node, it is necessary to attain all his children's subgoals although to attain a goal of an OR node, it is enough to attain, at least, one child's subgoal.

The attack-defense tree represents an extension of the attack tree. The main difference between them is that the attack tree introduces just the probable attacks the attacker can launch to achieve their global goal, whereas the attack-defense tree model both the possible attacks that the attacker can initiate and the countermeasures that the defender can take to face these attacks [10].

Since the aim of the attack-defense tree is to analyze an attack-defense scenario, it could be represented as a two-player game (attacker and defender) in order to reflect the mutual interaction between the attacker and the defender.

5.4 Attack-Defense Tree for VANET Availability

Depending on the attack tree model given in our previous article [11], in this section, we construct the attack-defense tree model (Fig. 5.1) for protecting the availability of vehicular ad hoc networks and their services. Our attack-defense tree contains the attacker subgoals presented in [11] and their corresponding countermeasures

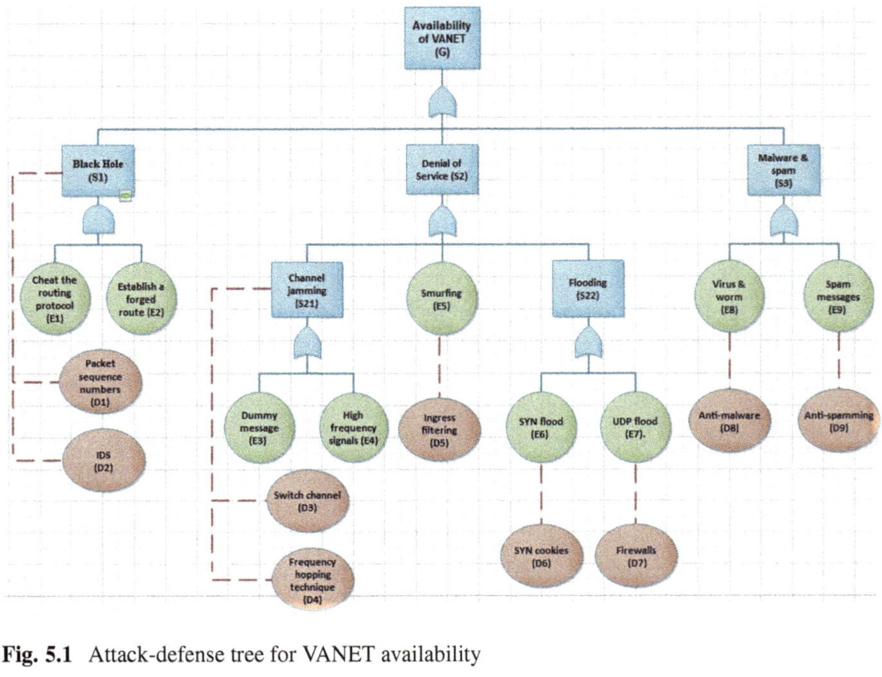

Fig. 5.1 Attack-defense tree for VANET availability

that represent the defender behavior against these attacks. However, the attacker subgoals are the following:

The black hole attack [12] denoted by (S1) is the first subgoal of the attacker. It is possible to accomplish this attack by merging two elements which are cheating the routing protocol and creating a forged path.

It could be achieved by combining two elements that cheat the routing protocol (E1) and establish a forged route (E2).

The second subgoal is the denial of service attack [13] referenced by (S2). This attack can be reached by channel jamming (S21) or smurfing attack (E5) or flooding attack (S22).

Transmitting dummy messages (E3) or broadcasting high-frequency signals (E4) can lead to channel jamming.

Flooding attack can be achieved through SYN flood (E6) or UDP flood (E7).

The final subgoal is malware and spam denoted by (S3). This subgoal can be attained by inserting viruses and worms in the wireless networks (E8) or through submitting spam messages (E9) to avoid broadcasting legitimate messages.

To combat these attacks, we adopt some security mechanisms illustrating the potential countermeasures used by the defender to protect the system [14].

To prevent VANET against black hole attack (S21), packet sequence numbers (D1) must be used in the packet header, or it is desired to implement an intrusion detection system (D2).

The usages of the packet sequence numbers (D1) in the packet header help the target node to detect the missing packet from the misplaced packet sequence number.

The intrusion detection system (IDS) (D2) represents a process that listens sneakily to the network traffic in order to identify unauthorized entry or to detect abnormal and suspicious activity. Moreover, this system verifies also data integrity. In VANET, IDS analyzes incoming and outgoing packets to detect malicious signatures. Sivaranjanadevi et al. [15] introduced an IDS using support vector machine technique and rough set theory in order to recognize malicious vehicles leading black hole attacks.

To defend VANET availability against DOS attack, we adopt the following countermeasures:

In order to protect the system against channel jamming (S21) attack, it is necessary to indicate to the OBU to switch channel (D3) or to employ the frequency hopping technique (D4).

To deal with the smurf attack (E5), we adopt the ingress filtering (D5) countermeasure. This countermeasure guarantees that incoming packets are actually coming from the networks they are pretending to come from. Network ingress filtering denied the attacking packages coming from a fake source address.

The SYN cookies (D6) can be used to safeguard the system against SYN flood attacks (E6). SYN cookies are specific values of the initial sequence numbers generated by a server (ISN, initial sequence number) during a TCP connection request.

Deployed firewalls (D7) in different network locations to filter out unwanted network traffic can deal with the UDP flood attack (E7). Nevertheless, the target node never gets malicious UDP packets and doesn't react to them because the firewall block them.

Anti-malware (D8) and anti-spamming (D9) should be utilized for protecting the network from malware (E8) and spam (E9).

The deployment of anti-malware (D8) in the roadside unit will defend vehicular networks from malwares.

The anti-spamming (D9) can block and minimize the undesirable effect of spam messages (E9) on the networks.

5.5 ROI and ROA for Attack-Defense Tree

This section evaluates the gains of the attacker from launching a specific attack and the gains of the defender from taking the corresponding countermeasures. Thus, we take into consideration the economic factors in the analysis of VANET availability. According to the attack-defense tree model, two costs are considered: the cost of attack and the cost of defense (security investment cost). These costs are used to define the Return on Investment (ROI) and Return on Attack (ROA) [16].

Actually, the Return on Investment (ROI) measures the benefit anticipated by the defender from an investment in security against the cost of implementation of defensive techniques. ROI is calculated by the following equation:

$$\mathrm{ROI} = \frac{\mathrm{ALE} \times RM - CI}{CI} \tag{5.1}$$

where ALE indicates the Annual Expected Loss due to VANET security threat; RM is the Risk Mitigation through the countermeasure, and it defines the countermeasure's efficiency in mitigating the risks of an attack against the system; and CI is the Cost of Investment that represents the cost to implement a certain countermeasure paid by the defender.

From the other side, the Return on Attack (ROA) measures the benefit anticipated by an attacker from a successful attack against the losses that he suffered because of the implementation of the security mechanism by his target. ROA is calculated by the following equation:

$$\mathrm{ROA} = \frac{GE \times (1 - RM) - (\mathrm{Cost}_A + \mathrm{Cost}_{AC})}{(\mathrm{Cost}_A + \mathrm{Cost}_{AC})} \tag{5.2}$$

where GE is the expected gain from a successful attack on a particular target, Cost_A is the cost sustained by the attacker to succeed, and Cost_{AC} is the supplementary cost caused by the countermeasure C implemented by the defender to mitigate the attack A.

Tables 5.1 and 5.2 represent the specific values chosen for all countermeasures and attacks, as well as the evaluation of the ROI and ROA, respectively.

Table 5.1 ROI evaluation

Attack	ALE	Countermeasures	RM	CI	ROI
S_1	8	D1 D2 D3, ..., D9	0.25 0.25 0	6 4 –	0.0128
S_{21}	8	D3 D4 D1, D2, D5, ..., D9	0.25 0.5 0	6 4 –	0.06
E_5	6	D5 D1, D2, D3, D4, D6, ..., D9	0.25 0	6 –	0.072
E_6	4	D6 D1, D2, D3, D4, D5, D7, D8, D9	0.5 0	6 –	0.102
E_7	6	D7 D1, D2, D3, D4, D5, D6, D8, D9	0.5 0	4 –	0.072
E_8	4	D8 D1, D2, D3, D4, D5, D6, D7, D9	0.5 0	8 –	0.072
E_9	4	D9 D1, D2, D3, D4, D5, D6, D7, D8	0.5 0	8 –	0.116

Table 5.2 ROI evaluation

Attack	GE	$Cost_A$	Countermeasures	$Cost_{AD}$	ROA
S_1	6	4	D1 D2 D3, ..., D9	2 4 0	−0.25 −0.437 0.5
S_{21}	6	6	D3 D4 D1, D2, D5, ..., D9	6 2 0	−0.625 −0.625 0
E_5	2	6	D5 D1, D2, D3, D4, D6, ..., D9	6 0	−0.875 −0.667
E_6	6	10	D6 D1, D2, D3, D4, D5, D7, D8, D9	2 0	−0.75 −0.4
E_7	8	10	D7 D1, D2, D3, D4, D5, D6, D8, D9	4 0	−0.714 −0.2
E_8	4	8	D8 D1, D2, D3, D4, D5, D6, D7, D9	4 0	−0.833 −0.5
E_9	4	8	D9 D1, D2, D3, D4, D5, D6, D7, D8	4 0	−0.833 −0.5

In Table 5.1, the reason why we chose the S1 subgoal instead of the E1 and E2 leaf nodes is related to the type of logical operator "AND" (shown in Fig. 5.1) of this subgoal of attack. In other words, the attacker will get no gain if he considers only one attack E1 or E2 since they must both execute to get to sub-objective S1. For the S21 sub-objective, we chose it instead of the leaf nodes E3 and E4 because both attacks can be defended by the same countermeasures D3 and D4. Moreover, the null values of RM mean that the countermeasure could not mitigate the attack and the dash "–" in the CI column means that the countermeasure has no effect on the result of its corresponding ROI. Furthermore, the ROI $= -1$ when a countermeasure could not mitigate an attack ($RM = 0$).

From Tables 5.1 and 5.2, we get the ROI and ROA values for each attack and countermeasure pair. The ROA and ROI values constitute the utility matrix of our attack-defense game introduced in the next section.

5.6 VANET Availability Attack-Defense Game

5.6.1 Basics of the Game Theory

Game theory plays an important role in theoretical and logical computer science to formalize semantics game or multi-agent systems [17]. It offers systematic analyses of the challenges presented by the strategic interaction of a number of rational actors who pursue their own objectives.

A game can be described as a situation in which at least one agent can act to maximize his gain by anticipating the responses of one or more other agents; the agents are called players. Players are generally assumed to be rational; they act for their own good and with respect for nature, social rules, etc. However, the game can be categorized by the number of players. Three major components are found in a complete game theory: the player, Nash equilibrium [18] (or strategies), and the utility function [19] (or payoff).

The player could be an individual (one player) or a team that contains several players (n players) with common benefits.

Nash equilibrium is regarded as a key component of the theory of games. It can be defined as a set of strategies that include only one strategy for each player. The main property in Nash equilibrium is that no player regrets his choice (he could not have done better by choosing another strategy) in view of the choice of others, the choices being, as always in game theory, simultaneous. So each player, in Nash equilibrium, correctly foresees the choice of the others and maximizes his gain, given this prediction.

The utility function (payoff) represents the level or the degree of satisfaction that the players derived from their events or strategies.

Nevertheless, the attack-defense tree could be represented as a game between two players (attacker and defender) to model attacking and defending strategies of the attacker and the defender, respectively. The next section models our attack-defense tree in order to illustrate the mutual interaction between the defender and the attacker.

5.6.2 Modeling VANET Availability Attack-Defense Game

This section analyzes the potential strategies of the attacker and of the defender using an attack-defense game for VANET availability based on our attack-defense tree. The attack-defense game is modeled as a static game where players take action whereas the action of the other players is unrecognized. Consequently, the attacker and the defender, in our attack-defense game, do not know the strategies of each other when they take actions. Moreover, we assume that players are rational in our attack-defense game model. This hypothesis maintains our analysis tractable while resolving the Nash equilibrium solution. Our game is defined as G (J, S, U), where J represents players, S represents a set of player's strategies, and U represents a set of utility functions.

Players The set of $J = \{ J_i\}$ contains the attacker J_1 and the defender J_2. Each player does not know the action that his opponent will play. For example, when the attacker wanted to inject a black hole attack ($S1$), the defender cannot receive any notification of this action that makes the game static.

Strategies Each player has a strategy S_j, where $j = 1$ or 2. In fact, all attacks are included in strategy S_1, and all countermeasures are included in strategy S_2. The attacks A_j that the attacker can select are {S1, S21, E5, E6, E7, E8, E9}, and the countermeasures that the defender can choose are $\{D_i \mid i = 1, \ldots, 9\}$.

Utility Functions The utility (or payoff) functions are described as follows: $U_1(D_i, A_j) =$ ROI (D_i, A_j) and $U_2(D_i, A_j) =$ ROA (D_i, A_j).

The utility matrix of the attack-defense game (ROI (D_i, A_j), ROA (D_i, A_j)) is shown in Table 5.3. In particular, Tables 5.1 and 5.2 illustrate parameters for calculating the defender's payoff (ROI) and the attacker's payoff (ROA), respectively.

It is helpful to adopt the dominated strategy [20] to optimize the method of finding the final solutions before resolving the game's Nash equilibrium. Nevertheless, in the case of non-zero-sum play, it is difficult to assume that one player is always looking to reduce the other's payoff even when it may be unfavorable to him. This can sometimes result in nonrational behavior (in the sense that the player does not choose the best action in the sense of his preferences). The notions of dominance and dominated strategies make it possible to correct this problem. They make it possible, based on the concept of rationality, to reduce the field of actions that will be analyzed, so it optimizes the method of finding the final solutions that lead to resolving the game's Nash equilibrium. Based on the dominated strategy concept, we can eliminate strategies that its payout is lower than any other strategy, according to the same strategy of the opponent in the attack-defense game.

Accordingly, we deduce, from Table 5.3, that the attacks E5, E6, E8, and E9 are dominated strategies for the attacker, and the countermeasures D1, D3, D5, D6, D8, and D9 represent the dominated strategies for the defender. Thus, the probability of selecting the attacks E5, E6, E8, or E9 or the countermeasures D1, D3, D5, D6, D8 or D9 is null for the attacker and the defender, respectively. Table 5.4 represents the reduction of the attack-defense game.

To calculate the probabilities that an attack or a countermeasure can reach, we adopt the following equations:

Table 5.3 The utility matrix

	S1	S21	E5	E6	E7	E8	E9
D1	−0.667,−0.25	−1, 0	−1,−0.667	−1, − 0.4	−1, − 0.2	−1, − 0.5	−1, − 0.5
D2	−0.5,−0.437	−1, 0	−1,−0.667	−1, − 0.4	−1, − 0.2	−1, − 0.5	−1, − 0.5
D3	−1, −0.5	− 0.667,− 0.625	−1,−0.667	−1, − 0.4	−1, − 0.2	−1, − 0.5	−1, − 0.5
D4	−1, − 0.5	0,− 0.625	−1,−0.667	−1, − 0.4	−1, − 0.2	−1, − 0.5	−1, − 0.5
D5	−1, − 0.5	−1, 0	−0.75,−0.875	−1, − 0.4	−1, − 0.2	−1, − 0.5	−1, − 0.5
D6	−1, − 0.5	−1, 0	−1,−0.667	−0.677,−0.75	−1, − 0.2	−1, − 0.5	−1, − 0.5
D7	−1, − 0.5	−1, 0	−1,−0.667	−1, − 0.4	−0.25,−0.714	−1, − 0.5	−1, − 0.5
D8	−1, − 0.5	−1, 0	−1,−0.667	−1, − 0.4	−1, − 0.2	−0.75,−0.833	−1, − 0.5
D9	−1, − 0.5	−1, 0	−1,−0.667	−1, − 0.4	−1, − 0.2	−1, − 0.5	−0.75,−0.833

Table 5.4 The attack-defense game reduction

	S1	S21	E7
D2	−0.5,−0.437	−1, 0	−1, − 0.2
D4	−1, − 0.5	0,−0.625	−1, − 0.2
D7	−1, − 0.5	−1, 0	−0.25,−0.714

$$\sum_{k=1}^{p} P_{D_{ik}}\text{ROA}\left(D_{ik}, A_{j1}\right) = \sum_{k=1}^{p} P_{D_{ik}}\text{ROA}\left(D_{ik}, A_{j2}\right)$$

$$= \ldots\ldots\ldots\ldots\ldots\ldots\ldots\ldots\ldots$$

$$= \sum_{k=1}^{p} P_{D_{ik}}\text{ROA}\left(D_{ik}, A_{jq}\right) \tag{5.3}$$

$$\sum_{k=1}^{q} P_{A_{jk}}\text{ROI}\left(D_{i1}, A_{jk}\right) = \sum_{k=1}^{q} P_{A_{jk}}\text{ROI}\left(D_{i2}, A_{jk}\right)$$

$$= \ldots\ldots\ldots\ldots\ldots\ldots\ldots\ldots\ldots$$

$$= \sum_{k=1}^{q} P_{A_{jk}}\text{ROI}\left(D_{ip}, A_{jk}\right) \tag{5.4}$$

where p is the number of countermeasures (D_{ik}, $k = 1, \ldots, p$) taken by the defender with non-null probabilities, and q is the number of attacks (A_{jk}, $k = 1, \ldots, q$) led by the attacker with non-null probabilities.

From the reduction of the attack-defense game represented in Table 5.4, we obtain the following equations:

$$-0,625\ \mathrm{P}_{\mathrm{D}_4} = 0,437\ \mathrm{P}_{\mathrm{D}_2} + 0,5\ \mathrm{P}_{\mathrm{D}_4} - 0,5\ \mathrm{P}_{\mathrm{D}_7}$$

$$= -0,2\ \mathrm{P}_{\mathrm{D}_2} - 0,2\ \mathrm{P}_{\mathrm{D}_4} - 0,714\ \mathrm{P}_{\mathrm{D}_7} \tag{5.5}$$

$$-0,5\ P_{S_1} - P_{s_{21}} - P_{E_7} = -P_{S_1} - P_{E_7}$$

$$= -P_{S_1} - P_{S_{21}} - 0,25\ P_{E_7} \tag{5.6}$$

From Eqs. 5.5 and 5.6, we can extract the probabilities of mixed strategies for the attacker and defender: $P_{S_1} = 0,45$, $P_{S_{21}} = 0,25$, $P_{E_7} = 0,30$, $\mathrm{P}_{\mathrm{D}_2} = 0,66$,

$P_{D_4} = 0,33$, and $P_{D_7} = 0,010$. However, we got the Nash equilibrium of this attack-defense game for the attacker who leads the attacks S1, S21, and E7 with the probabilities of 0.45, 0.25, and 0.30, respectively, and for the defender choosing the countermeasures D2, D4, and D7 with the probabilities of 0.66, 0.33, and 0.010, respectively.

The results of the final equilibrium reveal that the attacker will choose to carry out the black hole attack ($S1$) with the highest probability of 0.45, and the defender adopts the D2 countermeasure with a maximum probability of 0.66. Accordingly, we can deduce that the defender shall, as far as possible, cooperate with all other vehicles by using IDS to mitigate the black hole attack lead against VANET availability.

5.7 Conclusion

Even if VANET are emerging as a promising approach to improve road security, their deployment has considerably increased the risks associated with attacks against the various services offered by this network. In this article, we have proposed a new security assessment approach based on the attack-defense tree to identify the potential attacks that can be conducted against VANET availability as well as the associated countermeasures that can face these attacks. Moreover, we have introduced the Return on Attack (ROA) and the Return on Investment (ROI) to represent the potential gain from initiating an attack against VANET availability or adopting a countermeasure to protect the system. We also studied the potential strategies of the attacker and the defender by modeling it as an attack-defense game using game theory. In addition, we have used the concept of Nash equilibrium to describe the state of stability of the system, in order to construct strong defense mechanisms.

References

1. M. Houmer, M.L. Hasnaoui, An enhancement of greedy perimeter stateless routing protocol in VANET. Procedia Comput. Sci. **160**, 101–108 (2019)
2. B. Mokhtar, M. Azab, Survey on security issues in vehicular ad hoc networks. Alex. Eng. J. **54**(4), 1115–1126 (2015)
3. R. Shrestha, R. Bajracharya, S.Y Nam, Challenges of future VANET and cloud-based approaches: wireless communications and mobile computing. (2018)
4. S. Zeadally, R. Hunt, Y.S. Chen, A. Irwin, A. Hassan, Vehicular ad hoc networks (VANETS): status, results, and challenges. Telecommun. Syst. **50**(4), 217–241 (2012)
5. S. Parsons, M. Wooldridge, Game theory and decision theory in multi-agent systems. Auton. Agent. Multi-Agent Syst. **5**(3), 243–254 (2002)
6. M. Raya, J.P. Hubaux, Securing vehicular ad hoc networks. J. Comput. Secur. **15**(1), 39–68 (2007)

7. B. Kordy, S. Mauw, S. Radomirović, P. Schweitzer *Foundations of Attack–defense Trees: In International Workshop on Formal Aspects in Security and Trust* (Springer, Berlin, Heidelberg. 2010), pp. 80–95
8. B. Schneier, *Secrets and Lies: Digital Security in a Networked World* (John Wiley & Sons, 2015)
9. B. Schneier, Attack trees: Dr. Dobb's J. **24**(12), 21–29 (1999)
10. S. Bistarelli, M. Dall'Aglio, P. Peretti, Strategic games on defense trees. FAST **4691**, 1–15 (2007)
11. M. Houmer, M.L. Hasnaoui, A risk and security assessment of VANET availability using attack tree concept. Int. J. Electr. Comput. Eng. **10**(6), 2088–8708 (2020)
12. V. Bibhu, R. Kumar, B.S. Kumar, D.K. Singh, Performance analysis of black hole attack in VANET. Int. J. Comput. Netw. Inf. Secur. **4**(11), 47 (2012)
13. D. Rampaul, R.K. Patial, D. Kumar, Detection of DoS attack in VANETs. Indian J. Sci. Technol. **9**(47), 1–6 (2016)
14. M.S. Al-Kahtani, Survey on security attacks in vehicular ad hoc networks (VANETs).: *In the 6th International Conference on Signal Processing and Communication Systems*, IEEE. (2012), pp. 1–9
15. P. Sivaranjanadevi, M. Geetanjali, S. Balaganesh, T. Poongothai, An effective intrusion system for mobile ad hoc networks using rough set theory and support vector machine. IJCA Proc. EGovernance Cloud Comput. Serv. **2**, 1–7 (2012)
16. S. Bahamou, M.D. El Ouadghiri, J.M. Bonnin, When Game Theory Meets VANET's Security and Privacy, in *Proceedings of the 14th International Conference on Advances in Mobile Computing and Multi Media* (2016), pp. 292–297
17. A. González-Briones, P. Chamoso, S. Rodríguez, A. González-Arrieta, J.M. Corchado, Encouraging the recycling process of urban waste by means of game theory techniques using a multi-agent architecture, in *Ibero-American Conference on Artificial Intelligence* (Springer, Cham. 2018), pp. 120–131
18. E. Maskin, Nash equilibrium and welfare optimality. Rev. Econ. Stud. **66**(1), 23–38 (1999)
19. A.B. Leoneti, Utility function for modeling group multicriteria decision making problems as games. Operations Research Perspectives **3**, 21–26 (2016)
20. L. Wang, Y. Liu, Z. Wu, F.E. Alsaadi, Strategy optimization for static games based on STP method. Appl. Math. Comput. **316**, 390–399 (2018)

Chapter 6
A Secure Vehicle to Everything (V2X) Communication Model for Intelligent Transportation System

Mariya Ouaissa, Mariyam Ouaissa, Meriem Houmer, Sara El Hamdani, and Zakaria Boulouard

6.1 Introduction

Intelligent transportation systems (ITS) refer to the applications of modern information and communication technologies in the transportation area [1]. ITS are present in several activity domains, in optimizing the deployment of infrastructure transport; in improving safety, in particular road security; and in the development of services [2]. In addition, the use of intelligent transportations is a section of sustainable development, and these networks help to monitor mobility and to transfer the vehicle to an environmentally friendly mode. They are subject to tight economic competition on a global level [3]. More particularly, it concerns all the systems making it possible to collect, store, process, and distribute information relating to the movement of people and, in particular, traveler information systems, electronic payments, management of freight, collective transport fleet management, traffic management assistance, and driving assistance [4].

The automotive industry is going through a period of rapid technological change. This includes supporting intelligent transportation systems that will connect

M. Ouaissa (✉) · M. Ouaissa
Moulay Ismail University, Meknes, Morocco
e-mail: mariya.ouaissa@edu.umi.ac.ma; mariyam.ouaissa@edu.umi.ac.ma

M. Houmer
ENSAM Meknes, Moulay Ismail University, Meknes, Morocco

S. El Hamdani
Faculty of Science, Moulay Ismail University, Meknes, Morocco
e-mail: s.elhamdani@edu.umi.ac.ma

Z. Boulouard
Faculty of Sciences and Techniques Mohammedia, Hassan II University, Casablanca, Morocco
e-mail: zakaria.boulouard@fstm.ac.ma

M. Ouaissa et al. (eds.), *Computational Intelligence in Recent Communication Networks*, EAI/Springer Innovations in Communication and Computing,
https://doi.org/10.1007/978-3-030-77185-0_6

vehicles, road infrastructure, people, and wide area networks to enable safer and more efficient road transport. The application of Vehicle to Everything (V2X) systems in the transport sector makes it possible in particular to augment the roadway safety but also to rise the efficiency of public transport. These systems also help to enhance the traffic efficiency while reducing congestion and reducing the environmental impact of road transport [5].

The security of V2X communication is also a priority subject. Regulators have emphasized the need to protect messages from fraudulent or misleading use. They also require that the user's personal data be protected and preserved so that the movements of drivers cannot be easily tracked or reconstructed by listening to and compiling all the messages sent and received by a car over a given period [6].

In this article, we propose a lightweight and secure authentication scheme for V2X communication. Our new solution uses Attribute-based Signature (ABS) to achieve message authentication, privacy, and integrity and to help vehicles to securely receive a secret key and message from an RSU or other vehicles.

The rest of this work is organized as follows: The next section presents an overview of vehicular networks. Section 6.3 describes the V2X communication. We investigate the security requirement in Section 6.4, and we study the methods and mechanisms cryptographic used in our solution in Section 6.5. Section 6.6 proposes a lightweight authentication scheme for V2X communication. In section 6.7, we analyze the security and performance of our proposition. Finally, we draw our conclusion in Sect. 6.8.

6.2 Overview of Vehicular Networks

In recent years, continuous advancements in wireless communications technologies have sparked the emergence of a range of new networks aimed at expanding connectivity in areas that do not often deploy wired technologies. Vehicle networks, therefore, become a promising research field, which attracts further interest from academics, automakers, and telecommunications operators [7]. However, such networks allow vehicle communication, and we can hope that they will reduce the time spent on the roads, make it more user-friendly, and even more improve road safety. In what follows, we will present the description of the components, technologies, and communication architectures used in vehicular networks [8].

6.2.1 Vehicular Networks Components

The core elements that make up vehicular networks are [9, 10]:

- **Road Users:** In vehicular networks, drivers and passengers can participate in communications. These users may be the consumer or the supplier of content.

Today, although mobile and onboard equipment is used by vehicles, the amount of road accidents and traffic violations has risen. Hence, it is essential to supply a mechanism that significantly reduces the driver's interaction with the onboard equipment when sharing content. As an example, architecture has been proposed which consists of automatically sharing information detected by the sensors in vehicles.

- **Onboard Units (OBUs):** Vehicles are equipped with OBUs, allowing users to communicate and share content with other vehicles or infrastructure in the same area. Many projects in Europe and worldwide have, in recent years, demonstrated the viability and effectiveness of vehicle communications in several applications using prototypes and evaluations. Thus, the deployment rate of vehicles equipped with onboard units is expected to rise in the next 15 years.
- **Infrastructure:** Certain applications that are built for VANET are centralized and utilize infrastructure where existing. In general, roadside units (RSUs) and cellular networks are the infrastructures which could be utilized in vehicular networks. Besides, the RSUs provide short-range communications and, owing to their high costs, are poorly used. On the other side, cellular networks provide long-range communications often expensive.

6.2.2 *Communication Architectures*

Vehicular networks are categorized by different kind of architectures, depending on the way, followed by road users to access and share content, it's about: centralized, decentralized, and hybrid architectures [11, 12].

6.2.2.1 Centralized Architecture: Vehicle-to-Infrastructure Communication

As illustrated in Fig. 6.1, the central architecture presumes that even though vehicles are physically nearby, the users have to access a centralized server continuously that handles their interactions with other users. No direct contact between vehicles occurs in such an architecture. In the literature, this communication is known as Vehicle-to-Infrastructure (V2I). Vehicles communicate indirectly through existing infrastructure such as RSUs and cellular networks. To date, RSUs have been poorly deployed due to their high cost. In addition, cellular systems are overwhelmed and not reach all areas by increasing demand.

6.2.2.2 Distributed Architecture: Vehicle-to-Vehicle Communication

The distributed architecture demonstrated in Fig. 6.2 includes only opportunistic communications between cars. When a vehicle meets other vehicles nearby,

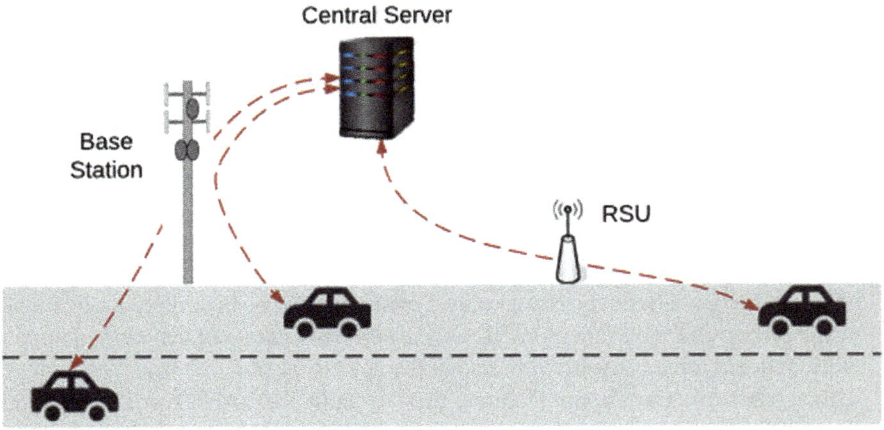

Fig. 6.1 Centralized architecture

Fig. 6.2 Distributed architecture

users can communicate and share content over the span of contacts between the Vehicle-to-Vehicle (V2V). V2V communication is inexpensive and offers significant transmission speed. On the other hand, this communication presents challenges, including low contact frequencies between low-density vehicles, short contact time due to speed and connection quality, and relay node selection.

6.2.2.3 Hybrid Architecture

The hybrid architecture, which we show in Fig. 6.3, includes both V2V communications and V2I communications. The infrastructure can be used optionally or when it is present. In areas where the infrastructure is nonexistent, this architecture opts for direct communications between vehicles.

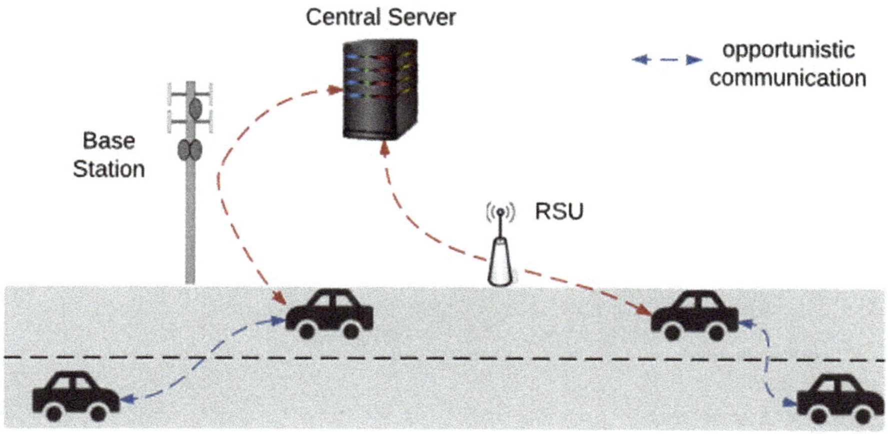

Fig. 6.3 Hybrid architecture

6.3 V2X Communications

The automotive world is about to experience the biggest revolution since its inception. Autonomous vehicles and intelligent transportation systems will transform the efficiency, comfort, safety, and environmental impact of road transport [13]. Mobile networks and technologies are at the heart of many of these advances, thanks to cellular connectivity supporting four basic types of use between the vehicle and everything (V2X) around it as shown in Fig. 6.4:

1. **Vehicle-to-Network (V2N):** connects vehicles to the mobile network to support services such as multimedia streaming for entertainment and connectivity for dynamic route management.
2. **Vehicle-to-Vehicle (V2V):** directly connects vehicles for early warning, including beyond the visual field, thus increasing the shorter range of the sensors on board.
3. **Vehicle-to-Infrastructure (V2I):** directly connects vehicles to road infrastructure such as traffic lights, which can also be connected to a larger mobile network.
4. **Vehicle-to-Pedestrian (V2P):** connects vehicles directly to pedestrians equipped with compatible mobile devices to issue alerts on potential surrounding dangers.

Currently, the most common type of V2X is V2N connectivity, which relies on mobile networks with permissions to use individual frequencies. A new variant of the Long-Term Evolution (LTE) standard called LTE-V2X was recently developed by third Generation Partnership Project (3GPP) in version 14 to support the connectivity of V2V, V2I, and V2P and will ensure the direct connection between devices (whether in vehicles, road infrastructure, or mobile devices) without the need for a network connection. The new LTE-V2X standard is a concurrent approach to a 10-year-old version of Wi-Fi called 802.11p that connects vehicles

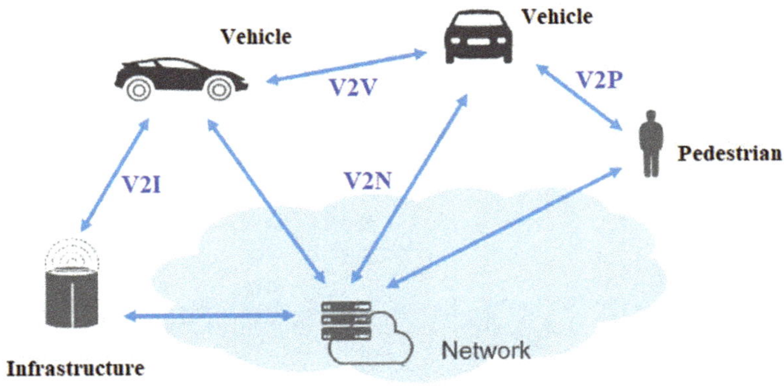

Fig. 6.4 V2X architecture

to infrastructure. Several different V2X standards have been developed based on 802.11p, including DSRC/WAVE and ITS-G5. These two regional variants of 802.11p are not compatible; therefore, they will require different equipment, which affects potential economies of scale, and the two have so far failed to make a significant market breakthrough. Some researchers have concluded that 802.11p solutions may be unsuitable for large-scale deployments. The new LTE-V2X standard complements the implementation of the existing V2N technology based on the LTE standard and offers considerable security advantages over 802.11p. Tests have shown that LTE-V2X technology supports approximately double the range of 802.11p, so it can transmit alerts to more vehicles or more reliable performance in the same range. Since it could take more than a decade before a significant number of vehicles, and road equipment are equipped with V2X capabilities, it is essential that there is a road map for technological development. There is already a clear road map for upgrading to LTE-V2X with full backward compatibility leading to Fifth Generation (5G) [14]. All vehicle manufacturers plan to support 5G-based V2X communication in the future; however, in the short term, there has been considerable support for 802.11p. This has changed since the advent of LTE-V2X, and the 5G Automotive Association (5GAA) whose 60 members include major vehicle manufacturers is currently supporting the LTE-V2X. In many countries around the world, regulators have reserved part of the spectrum for ITS, often in the 5.9 GHz band. This generally includes a portion dedicated to safety communications between vehicles, infrastructure, and pedestrians, where LTE-V2X and variants of 802.11p compete [15].

6.4 Security Requirements

It is essential that security requirements must always respect the proper functioning of a system in order to guarantee the security of the latter. When a requirement is not met, it presents a security problem. The requirements to be met by vehicular networks are authentication, confidentiality, non-repudiation, integrity, access control, requirements of real-time mode, and privacy protection. In the following, we detail these different requirements [16, 17].

6.4.1 Authentication

Authentication is a primary requirement of any system. For vehicular networks, it is necessary to know the information linked to the transmitting nodes such as its identifier, its geographical position, its address, and its properties. The main purpose of this requirement is to control the authorization levels of the vehicle in the network. Authentication helps to prevent attacks such as the Sybil Attack by specifying a unique identifier for each vehicle, and in this way, the latter cannot claim to have several identifiers in order to cause the network to malfunction. Several types of authentication have been presented in:

6.4.1.1 Authentication of the ID

A node must be able to identify the transmitters of a given message in a unique way. From this authentication, a transmitting vehicle can access the network.

6.4.1.2 Property Authentication

This type of authentication can determine whether the type of equipment in communication is another vehicle, an RSU, or other equipment.

6.4.2 Integrity

Integrity ensures that the message does not change between transmission and reception. The message receiver verifies the received message, making sure that the sender's identifier has not changed during transmission and that the message received is indeed the one that was sent. Integrity protects against modification, duplication, reorganization, and repetition of messages during transmission. Message integrity is managed by mechanisms that rely on one-way mathematical functions such as hashing functions and the message authentication code (MAC).

6.4.3 *Confidentiality*

The confidentiality of messages in vehicular networks depends on the application and the communication scenario. It allows the different nodes to have confidence in the messages broadcast in the network. There are two types of authentication: one for messages and the other for entities. Message authentication allows you to trace the source of the message, while that of the entities identifies the network nodes. Confidentiality can be implemented using the public/private keys for message encryption during communication. For example, in V2I communications, the RSUs and the vehicle share a session key after performing mutual authentication, so all messages are encrypted with the session key and attached with a message authentication code.

6.4.4 *Non-repudiation*

The purpose of non-repudiation is to collect, maintain, and make available all information related to the certainty of the entity disseminating messages, in order to avoid the harmful consequences that road safety applications can have on goods and people. Non-repudiation, therefore, depends on authentication. In this case, the implementation of the policy of non-repudiation in vehicular networks allows the system to identify the entity that disseminates a malicious message. For messages from safety and traffic management applications generally, the digital signature is used to guarantee non-repudiation. As for messages from comfort management applications, non-repudiation is not also necessary except for messages involving financial transactions.

6.4.5 *Availability*

The principle of availability is based on the fact that all entities on the network have permanent access to services or resources, so the services of road traffic management, safety, and comfort applications must always be available for vehicles. In order to ensure this permanence of services, vehicular networks must prevent denial of service attacks by using certain techniques such as frequency hopping and technology change.

6.4.6 *Access Control*

The role of access control is to determine rights and privileges in networks. Several applications differ according to the access levels granted to network entities. For example, traffic light control applications can be installed in police and emergency cars to facilitate their movement. In addition, certain communications such as those of the police or other authorities should not be listened to by other users. It is therefore essential to set up a system, which makes it possible to define all these access policies to guarantee access control in the network.

6.5 Preliminaries

This section explains the cryptographic methods applied in the proposed scheme, namely, Elliptic Curve Cryptography (ECC), Elliptic Curve Diffie-Hellman (ECDH) Protocol, and Attribute-based Signature (ABS) Algorithm.

6.5.1 *Encrypt Using Elliptical Curves*

Elliptical curves are well suited for public key cryptography. They make it possible to replace calculations on integers, or in groups Z/nZ, by calculations in groups associated with an elliptic curve [18].

6.5.1.1 Exchange of Keys by Elliptical Curves

Elliptic Curve Diffie-Hellman is a secret key exchange protocol. The description of the steps of algorithm are as follows [19]:

1. Alice and Bob want a common secret key. They agree on the choice of an elliptical curve E(K) where K is a finite body and on the choice of a point P of this curve. These choices are known to everyone.
2. They choose an integer a and an integer b, respectively. These integers will constitute their private keys.
3. Each of them then calculates, respectively, A = a.P and B = b.P.
4. Alice then sends Bob the point A, and the latter sends him the point B.
5. Alice then performs the calcul a.B = a.b.P and Bob the calcul b.A = a.b.P.

   They now have a commune a.b.P secret key.

6.5.1.2 Transmission of Messages

This time, we assume that Alice wants to send Bob a message using an elliptical curve encryption algorithm. Bob begins by making a public key as follows. It chooses an elliptical curve E (a, b, K), a point P of the curve, and an integer k_B. Its public key is constituted by the elliptic curve E (a, b, K) and the points P and $k_B.P$ of this elliptic curve. Its private key is the integer k_B, which cannot be found even knowing P and $k_B.P$, by the difficulty of solving the problem of the discrete logarithm on an elliptic curve.

When Alice wants to secretly send a point M on the elliptical curve to Bob, here is the exchange that takes place:

1. Alice reads Bob's public key (E, a, b, K), P and $k_B.P$.
2. Alice secretly and randomly chooses an integer n.
3. Alice calculates n.P and $M + n.k_B.P$ and sends these two points to Bob.
4. With his secret key k_B, Bob calculates $n.k_B.P$ from n.P, and then he calculates $(M + n.k_B.P) – n.k_B.P$. He, thus, finds M.

6.5.2 Attribute-Based Signature

Attribute-based Signature (ABS) is the most current variation of the fundamental cryptographic primitive, digital signature. ABS expands identity-based signatures where a signer has a number of attributes rather than a single string that corresponds to the identity of the signer. In ABS, the user gets the certificate from a certification authority called authority attribute for a series of attributes. An Attribute-based Signature guarantees to the verifier that the message is signed by a signatory whose number of attributes satisfies a complex predicate, and then the message is approved [20].

6.5.2.1 Computational Assumption

Suppose that G_1, G_2, and G_T are multiplicative groups of prime order p. Assume that $g_1 \in G_1$ and $g_2 \in G_2$ are primitives of G_1, G_2 where the bilinear pairing is the map e: G_1 x G_2 ➔ G_T containing the following characteristics [21]:

1. Bilinear: $e(g_1{}^a, g_2{}^b) = e(g_1, g_2)^{a*b}$ for every $g_1 \in G_1$, $g_2 \in G_2$, such that a, b ∈ Z.
2. Nondegenerate: There are $g_1 \in G_1$ and $g_2 \in G_2$ where $e(g_1, g_2) \neq 1$. However, a map is not passed the whole pairs in G_1 x G_2 to an identity in G_T.
3. Computability: There exists a cost-effective method to find $e(g_1, g_2)$ for any $g_1 \in G_1$ and $g_2 \in G_2$.

6.5.2.2 Lagrange Interpolation

Assume that q(x) the d-1 degree polynomial with every coefficient in Z_p. After that, provided one set of d points on a polynomial q(i) where $i \in f$ and f is a set of keys with $|f| = d$, we can utilize Lagrange interpolation to get q(j) for each $j \in Z_p$.

6.5.2.3 Attribute-Typed Signature

Each attribute-typed signature has the following algorithms:

1. **Setup Algorithm:** It is operated by the trusted agency. It accepts a security key d as an input and produces a set of public key (pk) 7-tuple and the corresponding master private key (sk).
2. **Key Generation Algorithm:** It is operated by a key generating agency. It accepts the master private key sk and the attributes set from a signer as an input and give k_l the signer secret key.
3. **Signing Algorithm:** It accepts a signer secret key k_l, the predicate l, and the document m to create the signature s.
4. **Verification Algorithm:** It checks that the signature s on the document m is true and when it is signed by the signer holds attributes which convince a predicate l.

6.6 Proposed Scheme

In this work, we propose a V2X communication security scheme, which consists, on the one hand, of a security phase without a trusted third-party authority, where each vehicle is authenticated with RSU of which it is part, and, on the other hand, a phase which guarantees V2V communication security for reliable data transmission. To achieve this, we have defined two objectives:

- First, ensure vehicle authentication via RSU by an ECDH algorithm as well as authenticate the message previously signed by the vehicle, using the digital signature algorithm based on attributes.
- Second, the transmission of data between vehicles that have successfully passed the message authentication step using the elliptical curve encryption algorithm.

The algorithms used are based on the concept of elliptic curves and attributes which gives them faster processing compared to their cryptographic counterparts. The proposed scheme ensures an aggregation of vehicle identification by the RSUs based on interaction areas without the intervention of a trusted third party (Figs. 6.5 and 6.6):

- The vehicle signals its presence in the interaction zone by sending a Beacon message, and it requests a shared secret.

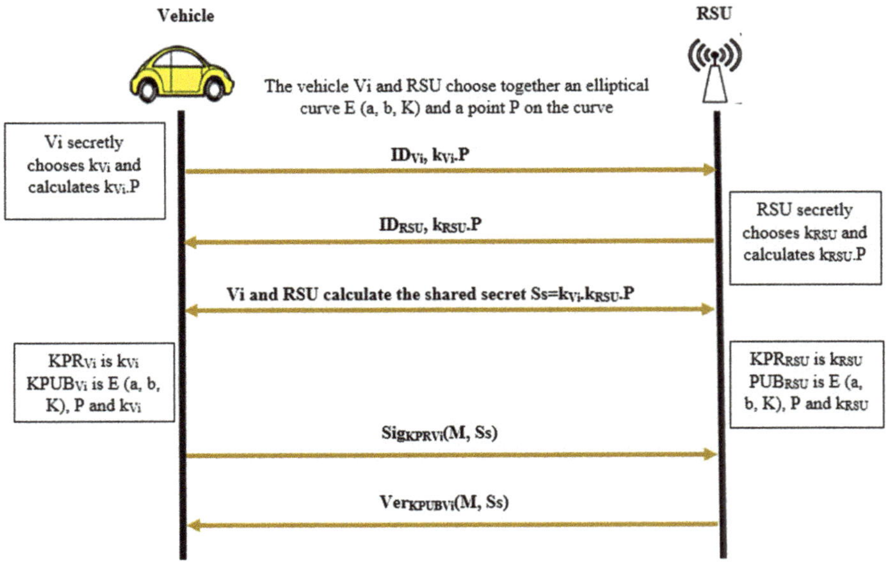

Fig. 6.5 Authentication and communication between vehicle and RSU

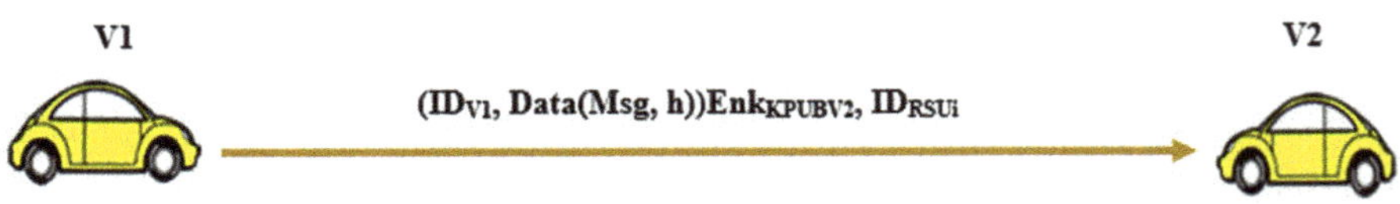

Fig. 6.6 Communication between vehicles

- The vehicle and the RSU concerned apply an ECDH protocol, and each sends its shared secret to the other to generate and exchange peers of cryptographic keys.
- The authentication of the vehicle by the RSU is ensured by the ABS signature process using the private key KPR and the ABS verification using the public key KPUB.

6.7 Validation and Evaluation

In the present section, we describe the verification tool used in our work, we analyze the security requirements and the formal verification of our model, and in addition, we evaluate the performance in terms of operational cost.

6.7.1 *Verification Environment*

In July 2005, the partners of the European AVISPA project published their work to develop a platform containing four protocol analysis tools and allowing the detection of logical attacks on security protocols. This platform also suggests improvements ensuring the validity of confidentiality and authentication properties. The verification techniques used by AVISPA [22] are techniques based on the principle of model checking. The four AVISPA tools are On-the-fly Model-Checker (OFMC), CL-based Attack Searcher (CL-AtSe), SAT-based Model-Checker (SATMC), and Tree-Automata-based Protocol Analyzer (TA4SP).

6.7.1.1 High-Level Protocol Specification Language (HLPSL)

High-Level Protocol Specification Language (HLPSL) [23] is a formal expressive specification language based on role descriptions. It supports different cryptographic primitives (symmetric and asymmetric keys, hash functions) and their algebraic properties (or exclusive, exponent). The purpose of these specifications is to be able to verify security properties, authentication, and confidentiality. The main idea is to represent a cryptographic protocol by a system of states/transitions for which it is possible to verify security properties expressed in linear temporal logic. The transitions define the behavior of the security protocol, and thus, from the initial state, we are able to enumerate the achievable states of the studied protocol. The HLPSL protocol specifications are divided into roles, which are divided into two distinct categories: basic roles and composition roles. The first type represents the agents participating in the protocols, while the second represents the scenarios of basic roles. At the end of the specification, the security properties to be checked are determined.

6.7.1.2 Verification Hypotheses

It is also important to model the intruder in the sense of the modeling of security protocols, i.e., to identify and minimize his behavior. The hypotheses used in this sense are collected under the name "Delev-Yao model" [24]. Two important assumptions are made in this model: perfect coding and network intruder. In particular, perfect encryption guarantees a message m encrypted with a key k can be decoded by an attacker only if it has the reverse of that key. The other assumption "the hacker is the network" means that the hacker can intercept and replace messages sent by honest protocol actors and send them messages under a false identity. Message transmissions in communication channels are public or private. In the canals, public messages exchanged are known by everyone, whether honest or not. But unlike, private channels are defined channels between certain honest participants. Consequently, an intruder cannot therefore listen to the messages, which circulate on this kind of channels.

6.7.1.3 Properties to Check

The following security properties should be checked:

- **Confidentiality:** It can also be called secret, verifying that the secret key is never transmitted in clear over the interface for the radio frequency that can be spied on.
- **Authentication Tag:** A reader must be able to check a proper authentication tag and to securely identify a tag.
- **Reader Authentication:** A tag needs to be confirmed if it is communicating with the correct reader.

6.7.2 Security Analysis

The primary objective of our proposal solution is to ensure that a secure exchange is possible and that authentication and integrity among vehicles and RSU are effective and also secure messages transmission between vehicles based on the different back end. In Figs. 6.7, 6.8, 6.9, and 6.10, we present, respectively, the authentication goals, the roles of vehicle 1, RSU, and vehicle 2.

After that the OFMC and CLAtSe back ends have run this specification, we may deduce that our scheme can resist to malicious attacks, such as Denial of Service (DoS), Man In The Middle (MITM), replay, and secrecy attacks and can accomplish our goals, firstly by ensuring the mutual authentication between vehicles and RSU where they exchange messages to authenticate each other and obtain a secret key. This secret sharing is carried out using the ECDH algorithm. Then the confidentiality of data, by encrypting them using the keys shared between vehicles and the signature, makes the non-repudiation service possible to verify the identity of the sender of this fact. Each message exchanged in the vehicular communication must be signed with the private key of the sender using ABS algorithm. The outputs of the model checking results are shown in Figs. 6.11 and 6.12.

Fig. 6.7 Analysis goals of our scheme

```
goal

authentication_on r1
authentication_on r2
authentication_on r3

end goal
```

```
role v1 (V1, RSU, V2 : agent, Ss: symmetric_key,
H : hash_func,
SND, RCV: channel (dy))
played_by Vi
def=
local
State: nat,
IDv1, IDv2, IDrsu, P, kv1, krsu, Mv1, Mv2: text,
Msg : message,
KPRv1, KPRv2, KPRrsu: private_key,
KPUBv1, KPUBv2, KPUBrsu: public_key
const
     r1, r2, r3 : protocol_id

init State := 1
transition
1. State = 1
/\SND (start) =|>
State' := 2
/\IDv1' := new()
/\IDrsu' := new()
/\Mv1' := (kv1.P)
/\Mv2' := (krsu.P)
/\SND (IDv1', Mv1')
/\RCV (IDrsu', Mv2')
2.  State = 2
/\SND (kv1'. krsu', P') =|>
State' := 3
/\Ss' := new()
/\Msg' :=new()
/\KPUBv1' := new()
/\KPRv1' := new()
/\SND ((Ss', Msg').KPRv1')
/\RCV ((Ss', Msg').KPUBv1')
3. State = 3
/\KPUBv2' := new()
/\SND ((IDv1', H(Msg')). KPUBv2', IDrsu)
 =|>
State' := 4
/\KPUBv1' := new()
/\IDv2' := new()
/\RCV ((IDv2', H(Msg')). KPUBv1', IDrsu)
end role
```

Fig. 6.8 Role of vehicle 1

```
role rsu (V1, RSU, V2 : agent, Ss: symmetric_key,
H : hash_func,
SND, RCV: channel (dy))
played_by Vi
def=
local
State: nat,
IDv1, IDv2, IDrsu, P, kv1, krsu, Mv1, Mv2: text,
Msg : message,
KPRv1, KPRv2, KPRrsu: private_key,
KPUBv1, KPUBv2, KPUBrsu: public_key
const
     r1, r2, r3 : protocol_id

init State := 1
transition
1. State = 0
/\RCV (start) =|>
State' := 1
/\IDv1' := new()
/\IDrsu' := new()
/\Mv1' := (kv1.P)
/\Mv2' := (krsu.P)
/\RCV (IDv1', Mv1')
/\SND (IDrsu', Mv2')
2.  State = 1
/\RCV (kv1'. krsu', P') =|>
State' := 2
/\Cv' := new()
/\SND (Cv', Nonce')
/\Ss' := new()
/\Msg' :=new()
/\KPRv1' := new()
/\KPUBv1' := new()
/\RCV ((Ss', Msg').KPRv1')
/\SND ((Ss', Msg').KPUBv1')
end role
```

Fig. 6.9 Role of RSU

```
role v2 (V1, RSU, V2 : agent, Ss: symmetric_key,
H : hash_func,
SND, RCV: channel (dy))
played_by Vi
def=
local
State: nat,
IDv1, IDv2, IDrsu, P, kv1, krsu, Mv1, Mv2: text,
Msg : message,
KPRv1, KPRv2, KPRrsu: private_key,
KPUBv1, KPUBv2, KPUBrsu: public_key
const
    r1, r2, r3 : protocol_id

init State := 3
transition
1. State = 1
/\KPUBv2' := new()
/\IDv1' := new()
/\Msg' := new()
/\IDrsu' := new()
/\RCV ((IDv1', H(Msg')). KPUBv2', IDrsu)
 =|>
State' := 2
/\KPUBv2' := new()
/\IDv2' := new()
/\SND ((IDv2', H(Msg')). KPUBv2', IDrsu)
end role
```

Fig. 6.10 Role of vehicle 2

6.7.3 Performance Evaluation

In this part, we analyze and evaluate the performance of our system in terms of operational cost, by comparing the execution time of algorithms and protocols used in our scheme and other existing ones, to prove that our considerations are faster and more efficient to follow the requirement and security needs of V2X communication. In this context, we choose to implement using Crypto++ Library [25]. The execution time values of such single operations were obtained by measurements running on a test platform with 2.1 GHz processor under Ubuntu. Table 6.1 illustrates the operations and their execution time.

```
% OFMC
% Version of 2006/02/13
SUMMARY
  SAFE
DETAILS
  BOUNDED_NUMBER_OF_SESSIONS
PROTOCOL
  /home/span/span/testsuite/results/proposed_Scheme.if
GOAL
  as_specified
BACKEND
  OFMC
COMMENTS
STATISTICS
  parseTime: 0.00s
  searchTime: 0.47s
  visitedNodes: 131 nodes
  depth: 10 plies
```

Fig. 6.11 Results reported by the OFMC back end

```
SUMMARY
  SAFE

DETAILS
  BOUNDED_NUMBER_OF_SESSIONS
  TYPED_MODEL

PROTOCOL
  /home/span/span/testsuite/results/proposed_Scheme.if

GOAL
  As Specified

BACKEND
  CL-AtSe

STATISTICS

  Analysed   : 1 states
  Reachable  : 1 states
  Translation: 0.09 seconds
  Computation: 0.00 seconds
```

Fig. 6.12 Results reported by the CL-AtSe back end

Table 6.1 Execution time of algorithms/protocols

Operations	Algorithms/protocols	Execution time (μs)
Key exchange	DH	2
	ECDH	1.29
Digital signature	DSA	2.47
	ECDSA	5.38
	ABS	1.2
Asymmetric encryption	RSA	6.28
	El-Gamal	5.97
	ECC	4.33

DH Diffie-Hellman, *ECDH* Elliptic-Curve *Diffie*-Hellman
DSA Digital Signature Algorithm, *ECDSA* Elliptic Curve Digital Signature Algorithm, *ABS* Attribute-based Signature
RSA Rivest-Shamir-Adleman, *ECC* Elliptic-Curve Cryptography

6.8 Conclusion

The V2X networks refer to the V2V and V2I wireless communication systems for improving road safety and traffic movement between vehicles and between vehicles and structures, respectively. Developed for the transport sector, the new generation V2X systems support the reduction of the number of road fatalities by including, for example, systems to warn drivers of an imminent risk of collision. Among the problems that have arisen in these networks is the problem of authentication and confidentiality of data transmitted between entities. Our scheme allows vehicles to establish secret key and authenticate in RSU in a reliable way using digital signature by ABS algorithm and also secure the communication between vehicles using the cryptographic keys obtained in the phase of key exchange by ECDH. Formal verification and security analysis demonstrate that the proposed scheme can provide robust security and achieve its design objectives. In addition, the comparison of execution time of several algorithms and protocols demonstrates that the choices adopted in our scheme are the most faster and more responsive to the security requirement for vehicular communications.

References

1. S. El Hamdani, N. Benamar, A comprehensive study of intelligent transportation system architectures for road congestion avoidance, in *International Symposium on Ubiquitous Networking*, (Springer, Cham, 2017), pp. 95–106
2. G. Dimitrakopoulos, P. Demestichas, Intelligent transportation systems. IEEE Veh. Technol. Mag. **5**(1), 77–84 (2010)
3. S. El Hamdani, N. Benamar, Autonomous traffic management: open issues and new directions, in *2018 International Conference on Selected Topics in Mobile and Wireless Networking (MoWNeT)*, IEEE, (2018), pp. 1–5

4. N. El Faouzi, H. Leung, A. Kurian, Data fusion in intelligent transportation systems: progress and challenges–a survey. Information Fusion **12**(1), 4–10 (2011)
5. J. Wang, Y. Shao, Y. Ge, R. Yu, A survey of vehicle to everything (V2X) testing. Sensors **19**(2), 334 (2019)
6. K. Bian, G. Zhang, L. Song, Security in use cases of vehicle-to-everything communications, in *2017 IEEE 86th Vehicular Technology Conference (VTC-Fall)*, (2017), pp. 1–5
7. M. Ouaissa, M. Ouaissa, M. Houmer, A. Rhattoy, Improved MAC design-based dynamic duty cycle for vehicular communications over M2M system, in *Data Analytics and Management*, (Springer, Singapore, 2021), pp. 111–120
8. H. Vahdat-Nejad, A. Ramazani, T. Mohammadi, W. Mansoor, A survey on context-aware vehicular network applications. Veh. Commun. **3**, 43–57 (2016)
9. O. Trullols-Cruces, J. Morillo-Pozo, J.M. Barcelo, J. Garcia-Vidal, A cooperative vehicular network framework, in *IEEE International Conference on Communications*, (2009), pp. 1–6
10. M. Houmer, M. Ouaissa, M. Ouaissa, M. Hasnaoui, SE-GPSR: secured and enhanced greedy perimeter stateless routing protocol for vehicular ad hoc networks. iJIM **14**(13), 48–64 (2020)
11. K. Zheng, L. Hou, H. Meng, Q. Zheng, N. Lu, L. Lei, Soft-defined heterogeneous vehicular network: architecture and challenges. IEEE Netw. **30**(4), 72–80 (2016)
12. S. Eichler, A security architecture concept for vehicular network nodes, in *6th International Conference on Information, Communications & Signal Processing*, (2007), pp. 1–5
13. M. Boban, A. Kousaridas, K. Manolakis, J. Eichinger, W. Xu, Connected roads of the future: use cases, requirements, and design considerations for vehicle-to-everything communications. IEEE Veh. Technol. Mag. **13**(3), 110–123 (2018)
14. M. Ouaissa, M. Houmer, M. Ouaissa, An Enhanced Authentication Protocol based Group for Vehicular Communications over 5G Networks, in *2020 3rd International Conference on Advanced Communication Technologies and Networking (CommNet)*, IEEE, (2020), pp. 1–8
15. H. Moustafa, Y. Zhang, *Vehicular Networks: Techniques, Standards, and Applications* (Auerbach publications, 2009)
16. M. El Zarki, S. Mehrotra, G. Tsudik, N. Venkatasubramanian, Security issues in a future vehicular network. European Wireless 2 (2002)
17. Z. Lu, G. Qu, Z. Liu, A survey on recent advances in vehicular network security, trust, and privacy. IEEE Trans. Intell. Transp. Syst. **20**(2), 760–776 (2018)
18. Y. Choi, D. Lee, J. Kim, J. Jung, J. Nam, D. Won, Security enhanced user authentication protocol for wireless sensor networks using elliptic curves cryptography. Sensors **14**(6), 10081–10106 (2014)
19. A.P. Fournaris, I. Zafeirakis, C. Koulamas, N. Sklavos, O. Koufopavlou, Designing efficient elliptic curve Diffie-Hellman accelerators for embedded systems, in *IEEE International Symposium on Circuits and Systems (ISCAS)*, (2015), pp. 2025–2028
20. J. Li, M.H. Au, W. Susilo, D. Xie, K. Ren, Attribute-based signature and its applications, in *5th ACM Symposium on Information, Computer and Communications Security*, (2010), pp. 60–69
21. J.A. Sattar, Security of attribute based signature scheme. Eur. J. Sci. Res. **123**(3), 296–303 (2014)
22. AVISPA Project. http://www.avispa-project.org/
23. D. Von Oheimb, The high-level protocol specification language HLPSL developed in the EU project AVISPA, in *APPSEM 2005 workshop*, (2005), pp. 1–17
24. D. Dolev, A. Yao, On the security of public key protocols. IEEE Trans. Inf. Theory **29**(2), 198–208 (1983)
25. Crypto++ Library: http://www.cryptopp.com/

Chapter 7
A Novel Unsupervised Learning Method for Intrusion Detection in Software-Defined Networks

Zakaria Abou El Houda, Abdelhakim Senhaji Hafid, and Lyes Khoukhi

7.1 Introduction

Internet has become an essential element to individuals and organizations to build online businesses, conduct online education, and improve human's ability to work and perform activities at any time and from anywhere. The increased dependency on the Internet is accompanied by numerous security and privacy risks (e.g., security issues with Internet of Things (IoT) devices and theft of sensitive data). The new emerging security threats have been increasing in sophistication and strength and are predicted to cost huge financial losses of about $20 Billion (USD) by 2021 [1] for both educational and business organizations. Also, the recent emergence of IoT botnets (e.g., Mirai botnet), as well as the rapid growth in the number of insecure IoT devices, with an estimation of 75 billion connected devices by the end of 2025 [2], can provide attackers with more sophisticated tools (e.g., Botnet as-a-service) to conduct large scale and devastating attacks. To ensure the security of networks, Internet service providers (ISPs) make use of firewalls to control/filter connections

Z. Abou El Houda (✉)
NRL, Department of Computer Science and Operational Research, University of Montreal, Montreal, QC, Canada

ICD, University of Technology of Troyes, Troyes, France
e-mail: zakaria.abou.el.houda@umontreal.ca

A. Senhaji Hafid
NRL, Department of Computer Science and Operational Research, University of Montreal, Montreal, QC, Canada
e-mail: ahafid@iro.umontreal.ca

L. Khoukhi
GREYC CNRS, ENSICAEN, Normandy University, Caen, France
e-mail: lyes.khoukhi@ensicaen.fr

M. Ouaissa et al. (eds.), *Computational Intelligence in Recent Communication Networks*, EAI/Springer Innovations in Communication and Computing,
https://doi.org/10.1007/978-3-030-77185-0_7

between local networks and the Internet. Also, multiple security enforcement mechanisms such as antivirus, access control, and data encryption are used to protect the network from suspicious activity. However, it has been shown that these security measures are not sufficient to fully protect the network against zero-day attacks.

To this end, in addition to these preventive security mechanisms, intrusion detection systems (IDSs) are used to effectively and timely secure the network against any type of suspicious/unauthorized activity that can cause collateral damage to either data integrity, data confidentiality, or data availability. IDSs play a key role in ensuring the security of the network. They can be categorized into two categories: (1) signature filtering-based intrusion detection systems (misuse-based detection, SFIDS) and (2) anomaly-based intrusion detection systems (A-IDS). SFIDSs detect network anomalies by using a pre-defined attack pattern/signature of well-known intrusions, while A-IDSs learn normal behaviors of activities and consider any deviation as an intrusion. SFIDSs are vulnerable to zero-day attacks and are limited by their need of new/up-to-date attack patterns, while A-IDSs suffer from high false positive rate.

Software-defined network (SDN) is a novel paradigm that leverages network programmability to solve the limitations of conventional networks. SDN provides new capabilities, through a logically centralized component, to cope with the new emerging security threats ranging from DDoS attacks to phishing to data leakage [3–11].

The recent emergence of machine learning and deep learning (ML/DL) techniques has achieved promising results in many fields [12–16]. In order to efficiently/timely handle intrusion detection, IDSs adopted ML/DL techniques. ML/DL-based IDSs can effectively detect existing and new network security threats [17–19]. In this chapter, we investigate a novel problem of using unsupervised learning in the task of network anomaly/intrusion detection in software-defined networks (SDN). Most existing unsupervised ML/DL-based IDSs, such as clustering-based techniques (e.g., K-means [20]), try to find a profile of similar/normal data samples and then classify other/dissimilar samples as anomalies/intrusions. These techniques have two main drawbacks: (1) high false positive rates since they consider any deviation from the normal behavior as an intrusion and (2) high computational complexity since they focus on memorizing a large number of normal data sample patterns (i.e., hidden feature learning). To alleviate these issues, we develop a novel outlier detection method with Isolation Forest (IDS-IF) that can effectively detect network anomalies in SDN while having a low false positive rate as well as low computational complexity. To achieve this, IDS-IF isolates intrusions instead of profiling normal data samples. Anomalies/intrusions are mostly few and rare/different data samples (i.e., minority in the dataset); this makes them susceptible to isolation with low computation rather than profiling a large number of normal data samples. IDS-IF not only enhances the detection performance but also reduces the false positive rate as well as computational complexity. The experimental results, using the well-known public network security dataset KDD [21], show that IDS-IF outperforms the recent state-of-the-art outlier detection method (i.e., Local Outlier Factor (LOF)) in terms of accuracy, F1 score, and false positives rates, making it a promising method to cope with the new emerging security threats in SDN.

The main contributions of this chapter can be summarized as follows:

- We propose a novel light-weight unsupervised learning method for anomaly/intrusion detection in software-defined networks (SDN).
- We propose a novel outlier detection method with Isolation Forest (IDS-IF) to effectively detect network anomalies in SDN.
- We evaluate the performance of IDS-IF in terms of accuracy, F1 score, and computational complexity. We compare the performance of IDS-IF with three recent state-of-the-art outlier detection methods. The experimental results show that IDS-IF achieves security and high accuracy/F1 score in detecting new security threats, making it a promising method to cope with the new emerging security threats in SDN.

The remainder of this chapter is organized as follows. In Sect. 7.2, we present a review of related work in supervised and unsupervised ML/DL techniques for anomaly/intrusion detection. Then, we present a general overview of IDS-IF in Sect. 7.3. In Sect. 7.4, we describe the design and specification of IDS-IF. In Sect. 7.5, we present the performance evaluation of IDS-IF. Finally, Sect. 7.6 concludes the chapter.

7.2 Related Work

The new emerging security threats are becoming more devastating. Several state-of-the-art contributions have integrated supervised and unsupervised ML/DL techniques to improve the efficiency of traditional IDSs to cope with these attacks. In the following, we overview the most representative ML/DL-based IDSs as well as their security issues.

Ruoning et al. [22] proposed a novel real-time intrusion detection scheme that uses a dynamic cumulative-distance anomaly detection algorithm (i.e., k-nearest neighbors (k-NN)). The proposed scheme consists of a distributed data processing platform that uses flume [23] for a reliable log data aggregation and collection, and storm [24] for a distributed and reliable and stream processing. The effectiveness of the proposed scheme was evaluated using a real-world dataset. The experimental results did show that this algorithm is suitable for real-time network anomaly detection in high-speed networks.

Yang et al. [25] developed a new framework that uses a support vector machines (SVM) method to detect and mitigate network anomalies in SDN. This framework consists of three modules: (1) a traffic collection module to extract network traffic features/characteristics and prepare them for network traffic identification module; (2) a network anomaly identification module to perform the classification and to identify anomalies using the SVM method; and (3) a flow table delivery module to dynamically adjust OpenFlow (OF) rules according to the attack identification module. The effectiveness of the proposed framework was tested and evaluated using KDD'99 dataset.

Majjed et al. [26] designed an effective DL framework, self-taught learning (STL-IDS), that uses a sparse autoencoder (SAE) along with a support vector machines (SVM) method to detect and mitigate network anomalies. STL-IDS uses a feature selection method and a dimensionality reduction scheme to reduce training time complexity while improving the prediction accuracy using the SVM algorithm.

Taher et al. [27] proposed a novel supervised ML method that uses artificial neural network (ANN) along with a feature selection method to detect and mitigate network anomalies. The authors have shown that ANN with feature selection method outperforms SVM with respect to intrusion detection rate. The effectiveness of the proposed framework was tested and evaluated using NSL-KDD dataset.

Yin et al. [28] proposed a novel deep learning scheme (RNN-IDS) that uses recurrent neural networks for network intrusion detection. The authors have studied the performance of RNN-IDS in binary and multi-class classification using NSL-KDD dataset. The experimental results did show that RNN-IDS outperforms shallow ML models with respect to detection accuracy.

Wang et al. [29] proposed a hierarchical spatial-temporal feature-based IDS called HAST-IDS. HAST-IDS has two main stages: (1) it uses deep convolutional neural networks (CNNs) to learn the low-level spatial features of network traffic and (2) it uses LSTM (long short-term memory) to learn a high-level temporal feature. The effectiveness of the proposed framework was tested and evaluated using the standard DARPA and ISCX2012 datasets.

Tuan et al. [30] proposed a light-weight unsupervised learning scheme based on a Local Outlier Factor (LOF) algorithm to detect and mitigate network anomalies (e.g., DDoS attacks) in SDN. The LOF algorithm measures the local deviation for a given data sample with respect to its neighbors (i.e., local density). The proposed solution requires minimal network resources; its evaluation, using CAIDA dataset, did show promising result.

Gao et al. [31] proposed an adaptive ensemble learning method that combines multiple shallow ML models (i.e., decision trees (DT), support vector machine (SVM), logical regression (LR), k-nearest neighbors (KNN), AdaBoost, random forests (RF), and deep neural networks) to increase the detection rate of shallow ML models. Also, they have proposed an ensemble adaptive voting algorithm. The effectiveness of the proposed framework was tested and evaluated using NSL-KDD dataset.

Based on our analysis of existing contributions [22–31], we found that some of these schemes [27–29] are computationally expensive. Also, most of them suffer from high false positive rate since they consider any deviation from the normal behavior as an intrusion. To address the shortcomings of existing solutions [22–31], we propose a novel outlier detection method with Isolation Forest (IDS-IF) to effectively detect network anomalies in SDN. IDS-IF isolates intrusions instead of profiling normal data samples; it does not use any computationally expensive method (i.e., density measure and distance measure) to detect intrusions. Also,

IDS-IF can handle large size and extremely high-dimensional problems. IDS-IF not only enhances the detection performance but also reduces false positive rate as well as computational complexity. The experimental results, using the well-known public network security dataset KDD [21], show that IDS-IF outperforms the recent state-of-the-art outlier detection method [30] in terms of accuracy, F1 score, and false positives rates, making it a promising method to cope with the new emerging security threats in SDN.

7.3 IDS-IF: An Overview

When designing IDS-IF, we did consider the following goals/objectives. First, IDS-IF should ensure/guarantee a full protection from the new emerging security threats. Unlike existing ML/DL-based IDSs, [22–31] that try to find a profile of similar/normal data samples then classify other/dissimilar samples as anomalies/intrusions. IDS-IF aims to isolate intrusions instead of profiling normal data samples. Anomalies/intrusions are mostly few and rare/different data samples that make them susceptible to isolation with low computation rather than profiling normal data samples that consist of a large number of data samples. Second, these anomalies/intrusions should be effectively and timely detected/mitigated, using an OpenFlow (OF) security policy, and the overall system has to be as secure as possible.

Figure 7.1 shows the architecture of IDS-IF. IDS-IF has two phases: (1) a novel outlier detection method with Isolation Forest (IDS-IF) to effectively detect network anomalies in SDN; this method is implemented in the application plane (i.e., top of the SDN controller); and (2) a security policy mitigation scheme to timely and effectively detect and mitigate these network security threats. The Northbound API (i.e., REST API) is used in the detection/mitigation process to offer the inter-operability to use/manage any type of SDN controller (e.g., Ryu OpenFlow Controller [32], Floodlight OpenFlow Controller [33]).

7.4 IDS-IF

In this section, we describe in more detail IDS-IF; in particular, we describe how it effectively isolates anomalies without profiling normal data sample.

IDS-IF isolates intrusions/anomalies instead of profiling normal data samples; it uses the fact that anomalies/intrusions are mostly few and rare/different data samples, which make them susceptible to isolation with low computation rather than profiling a large number of normal data samples. IDS-IF is mainly based on decisions trees; we define path length $p(z)$ of a data point z as the number of edges that the data point z traverses from the root of the tree to the external node of the tree. Abnormal data points are mostly few and have different representations (e.g.,

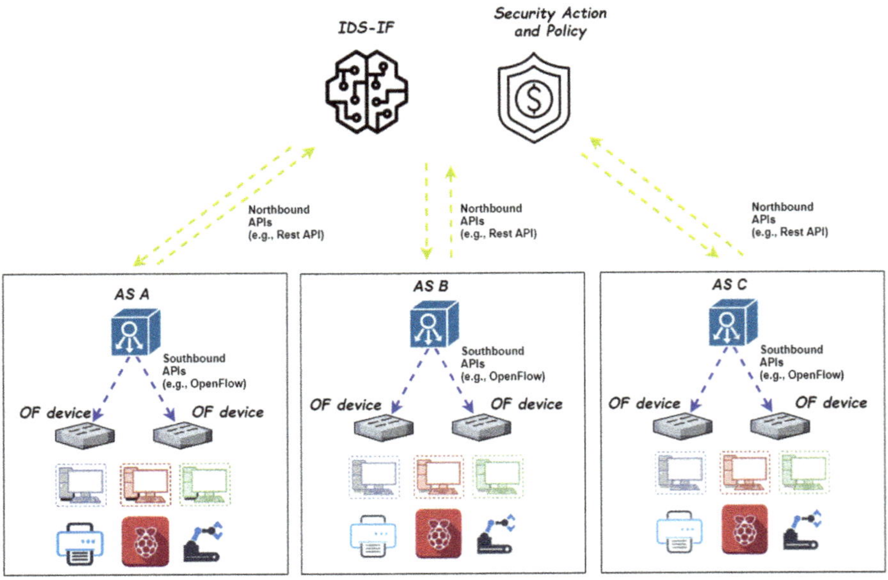

Fig. 7.1 System architecture

feature space) resulting in a shorter path $p(z)$ in the tree. The closer the data points to the root of the tree (i.e., shorter path length $p(z)$), the more likely these data points are intrusions.

In IDS-IF, a forest of trees is constructed using multiple decision trees; each tree trains on a sub-set of data. To construct an optimal tree, we select a random subset of features. Then, we construct multiple splits using different features. Several methods (e.g., Information Gain (IG) using entropy and Gini Index) can be used to select the best feature that maximizes the information for a particular split in the tree. In IG, we use entropy to measure the degree of randomness; the higher the value of IG, the more the order of disorder. Thus, the intention is to decrease the entropy from the top of the tree (i.e., root node) to the bottom of the tree (i.e., leaf nodes). IG is defined as follows:

$$\mathcal{IG} = -\sum_{j=1}^{N} p_j * log(p_j), \tag{7.1}$$

where p_j is the class probability; it is the proportion of class j in the dataset. Gini Index or Gini impurity computes the amount of probability of a particular feature that is misclassified when selected randomly. Gini Index measure is defined as follows:

$$\mathcal{GI} = 1 - \sum_{j=1}^{N} {q_j}^2, \tag{7.2}$$

where q_j is the probability of a misclassified feature.

Once an optimal tree is constructed using the aforementioned metrics, we compute an anomaly score, for each data point z, as follows:

$$sc(z, m) = 2^{\frac{-E(p(z))}{c(m)}}, \tag{7.3}$$

where $p(z)$ is the path length of data point z, m is the number of external nodes, $E(p(z))$ is the average of $p(z)$, and $c(m)$ is the average path length of unsuccessful searches in a binary search tree (BST); it is defined as follows:

$$c(m) = 2H(m-1) - (2\frac{m-1}{m}), \tag{7.4}$$

where $H(j)$ is an harmonic number that can be estimated by Euler's constant (i.e., $ln(j) + 0.5772156649$).

Using the anomaly score in Eq. 7.3, we can classify each data point z as either normal or abnormal. A score s close to 1 for a particular data point z indicates that z is very likely an abnormal data point, while a score s close to 0 indicates that z is very likely a normal data point.

7.5 Evaluation of IDS-IF

In this section, we present the evaluation of IDS-IF. First, we introduce the experimental environment. Then, we evaluate the performance of IDS-IF.

7.5.1 Experimental Environment

The implementation of IDS-IF is done using scikit-learn [34], an open-source library that integrates a wide range of supervised and unsupervised machine learning techniques. We used various functions of scikit-learn to implement IDS-IF. We run our experiments on Google Colaboratory [35] using the Tesla T4 GPU. We used the well-known public network security dataset KDD'99 dataset [21] that was created and maintained by MIT Lincoln Laboratory for 1998 DARPA Intrusion Detection Evaluation Program. KDD'99 data records were collected over 9 weeks during the Third International Knowledge Discovery and Data Mining Competition, in a local area network (LAN) that simulates a typical United States Air Force LAN. It contains about five million connection records. The dataset was created

Sample content	**type**
0, tcp, http, SF, 181, 5450, 0, 0, 0, 0, 0, 1, 0, 0, 0, 0, 0, 0, 0, 0, 0, 8, 8, 0, 0, 0, 0, 1, 0, 0, 9, 9, 1, 0, 0.11, 0, 0, 0, 0, 0	**normal**
0, tcp, private, S0, 0, 0, 0, 0, 0, 0, 0, 0, 0, 0, 0, 0, 0, 0, 0, 0, 0, 143, 15, 1, 1, 0, 0, 0.1, 0.06, 0, 255, 15, 0.06, 0.06, 0, 0, 1, 1, 0, 0, neptune	**DoS**
0, udp, private, SF, 207, 0, 0, 0, 0, 0, 0, 0, 0, 0, 0, 0, 0, 0, 0, 0, 0, 2, 2, 0, 0, 0, 0, 1, 0, 0, 242, 202, 0.83, 0.01, 0.83, 0, 0, 0, 0, 0, nmap	**Probe**
0, tcp, telnet, SF, 129, 174, 0, 0, 0, 0, 1, 0, 0, 0, 0, 0, 0, 0, 0, 0, 0, 1, 1, 0, 0, 0, 0, 1.00, 0, 0, 255, 255, 1, 0, 0, 0, 0.01, 0.01, 0.02, 0.02, guess_passwd	**R2L**
8, tcp, ftp_data, SF, 613, 0, 0, 0, 0, 0, 0, 1, 0, 0, 0, 0, 0, 0, 0, 0, 0, 1, 1, 0, 0, 0, 0, 1, 0, 1, 1, 84, 1, 0, 1, 0.02, 0, 0, 0, 0, ftp_write	**U2R**

Fig. 7.2 Samples of KDD dataset

Table 7.1 Details of KDD, SA, and SF datasets

Dataset	Records	Dimensionality	Types of features	Label
KDD	4,898,431	41	Continuous/Discrete	Normal/Abnormal
SA	976,158	41	Continuous/Discrete	Normal/Abnormal
SF	699,691	4	Continuous/Discrete	Normal/Abnormal

with the aim of providing the research community with a large training dataset for supervised learning techniques. Figure 7.2 shows samples of the KDD dataset. Each data sample can be categorized as one of the five classes: Distributed Denial of Service (DDoS), Probe (Probing), User to Root Attack (U2R), and Remote to Local Attack (R2L).

KDD contains a large proportion of about 80% of abnormal data; this is unrealistic in real world. Thus, we transform KDD into two datasets: SA and SF. SA contains all normal observations with a small proportion of about 1% of abnormal data of KDD dataset. SF contains all normal observations with a small proportion of about 0.3% of abnormal data of KDD dataset. The objective is to produce data that is suitable for such unsupervised tasks. Table 7.1 shows the details of KDD, SA, and SF datasets. KDD dataset contains about five million of data records, and it contains 41 features per record that can be either continuous or discrete. SA dataset contains about one million of data records; it contains 41 features per record that can be either continuous or discrete. SF dataset contains about one million of data records; it contains four features per record that can be either continuous or discrete. The output label that contains the name of the attack is changed to a numerical value (see Table 7.2). For data pre-processing phase, we have encoded categorical/non-numeric features (i.e., flag feature and service feature) into numeric values. Then, we have re-scaled the values of the features according to Eq. (7.5) using a standardization technique:

$$X_i' = \frac{X_i - Mean(X_i)}{stdev(X_i)}, \tag{7.5}$$

Table 7.2 KDD labels

Label	Binary
Normal	0
DoS	1
Probe	1
R2L	1
U2R	1

Table 7.3 Confusion matrix

	Classified as abnormal	Classified as normal
Abnormal data	TP (True Positives)	FN (False Negatives)
Normal data	FP (False Positives)	TN (True Negatives)

where X_i denotes a data input feature (e.g., flag), $Mean(X_i)$ and $stdev(X_i)$ denote, respectively, the mean and standard deviation values of X_i.

7.5.2 *Performance Evaluation*

We evaluate the performance of IDS-IF in terms of accuracy, precision, detection rate (DR), area under the ROC curve (AUC), and F1 score. The ROC curve shows TPR (True Positive Rate) according to FPR (False Positive Rate). Finally, we use a confusion matrix to show the overall performance of IDS-IF (see Table 7.3).

TPs (True Positives) represent anomalies/intrusions that are correctly identified/classified as intrusions, FNs (False Negatives) represent anomalies/intrusions that are classified as normal data samples, FPs (False Positives) represent normal data samples that are identified/classified as anomalies/intrusions, and TNs (True Negatives) represent normal data samples that are identified/classified as normal data samples.

Accuracy is the percentage of the number of correct classifications over the total number of classifications. Accuracy is defined as follows:

$$\text{Accuracy} = \frac{TP + TN}{TP + TN + FP + FN}. \tag{7.6}$$

Precision is the percentage of the number of correct classifications of anomalies/intrusions over the total number of classified anomalies/intrusions. Precision is a good performance metric suited for a highly imbalanced class distribution; it is defined as follows:

$$\text{Precision} = \frac{TP}{TP + FP}. \tag{7.7}$$

Detection rate (DR) is the percentage of the number of correct classifications of anomalies/intrusions over the total number of presented anomalies/intrusions. DR is defined as follows:

$$Recall = DR = TPR = \frac{TP}{TP + FN}. \tag{7.8}$$

It is often convenient to combine multiple performance metrics (i.e., precision and recall) into a single metric called F1 score. F1 is the harmonic mean of the precision and recall; it is defined as follows:

$$F1 = \frac{2}{\frac{1}{Precision} + \frac{1}{Recall}}. \tag{7.9}$$

False positive rate (FPR) is the percentage of the number of anomalies/intrusions incorrectly classified as normal data over the total number of negative samples. FPR is defined as follows:

$$FPR = \frac{FP}{TN + FP}. \tag{7.10}$$

To test the effectiveness of IDS-IF, we defined two scenarios that are summarized in Table 7.4. In scenario 1, we consider a binary classification using the SA dataset that contains 1% of abnormal data of the KDD dataset. In scenario 2, we consider a binary classification using SF dataset that contains 0.3% of abnormal data of the KDD dataset. We compare the performance of IDS-IF with a recent state-of-the-art outlier detection method [30] in terms of Accuracy, precision, recall, and F1 score in both scenarios. Higher these permanence metrics values indicate a better classification model.

Figures 7.3 and 7.4 show the confusion matrices on the KDD dataset for scenario 1 and scenario 2, respectively. In scenario 1, IDS-IF achieves 89%, 97%, 88%, 94%, 92% in accuracy, precision, recall, AUC, and F1 score, respectively, compared to 81%, 93%, 82%, 46%, 87% in accuracy, precision, recall, AUC, and F1 score achieved by LOF, respectively. In scenario 2, IDS-IF achieves 89%, 96%, 88%, 85%, 91% in accuracy, precision, recall, AUC, and F1 score, respectively, compared to 81%, 91%, 81%, 47%, 87% in accuracy, precision, recall, AUC, and F1 score achieved by LOF, respectively. Table 7.5 shows the performance metrics of IDS-IF and LOF on the KDD dataset. We observe that, in both scenarios, IDS-IF achieves the highest accuracy, precision, recall, AUC, and F1 score, making it a promising method to mitigate the new emerging threats in SDN.

Table 7.4 Used scenarios

Scenarios	Type	Dataset
1	Binary classification	SA
2	Binary classification	SF

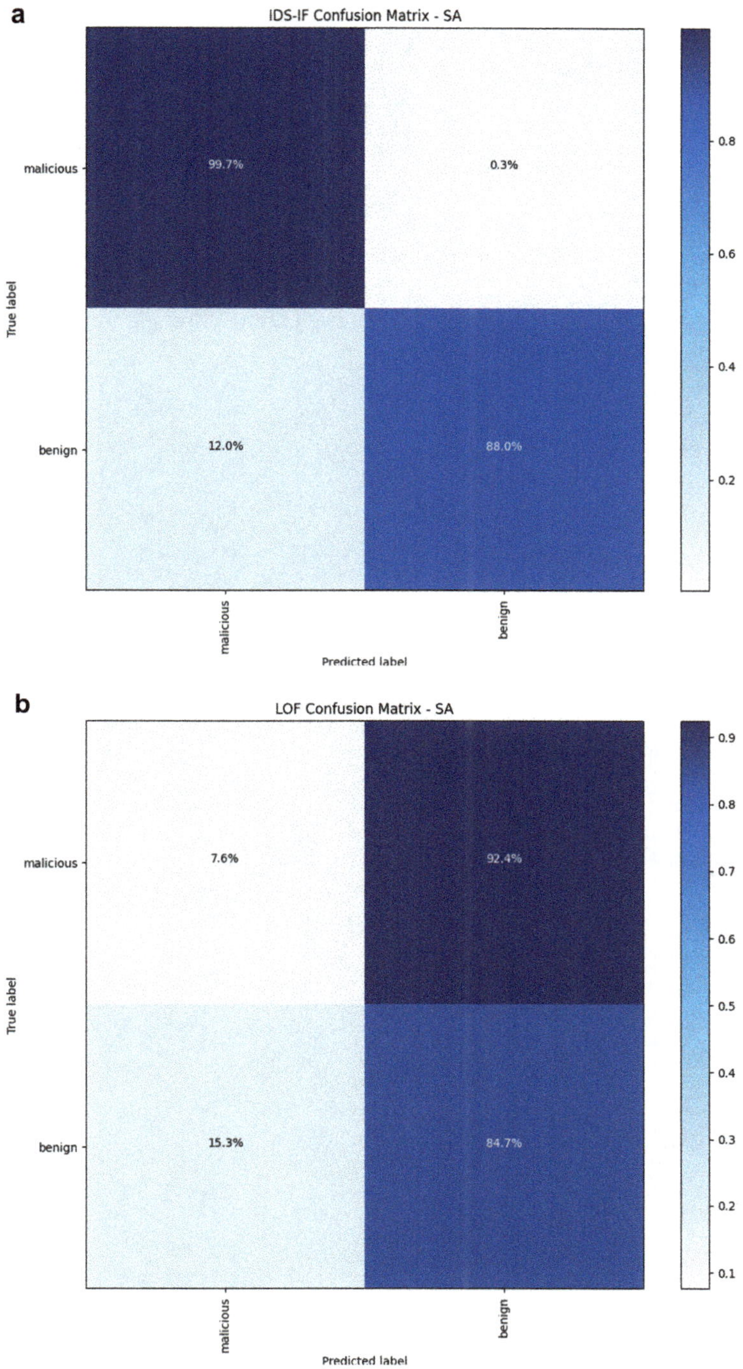

Fig. 7.3 Confusion matrices for scenario 1 on the KDD dataset for: (**a**) IDS-IF and (**b**) LOF

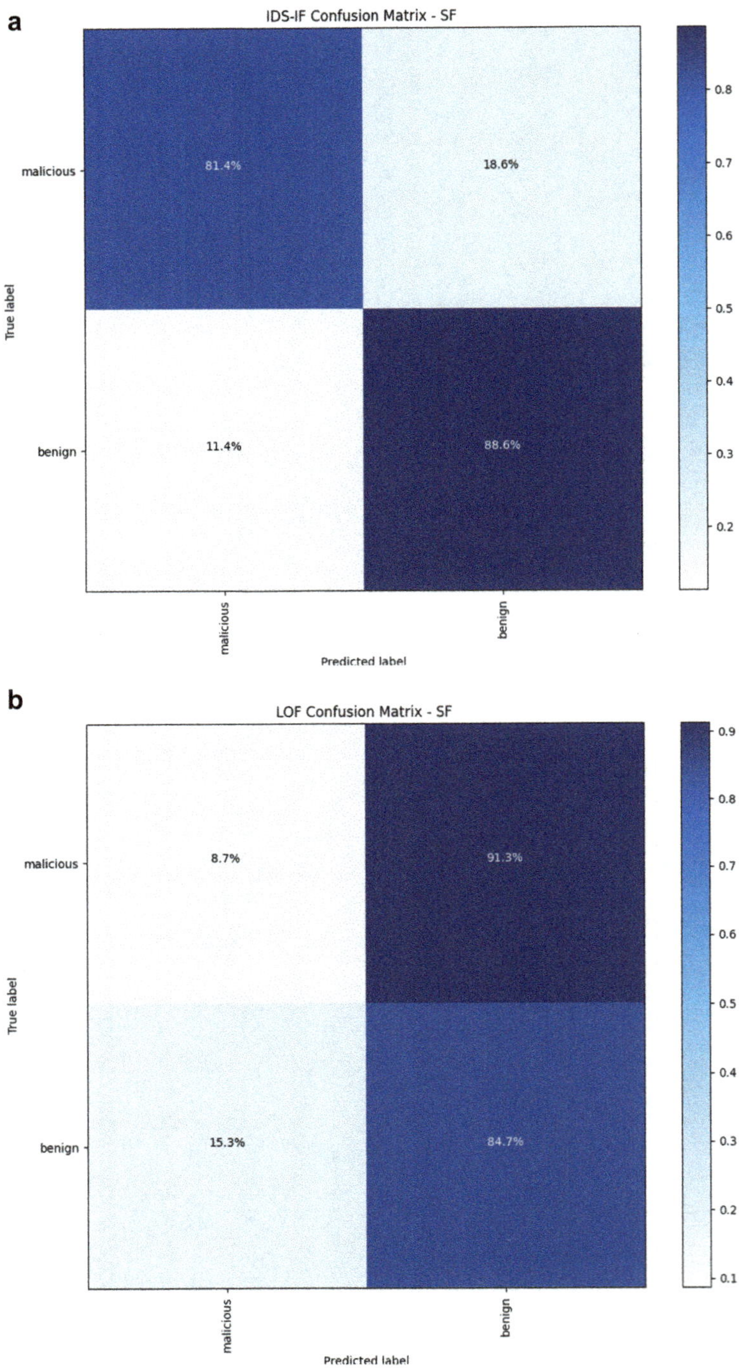

Fig. 7.4 Confusion matrices for scenario 2 on the KDD dataset for: (**a**) IDS-IF and (**b**) LOF

Table 7.5 Performance metrics of IDS-IF and LOF on *KDD* dataset

Methods	Scenarios	Accuracy	Precision	Recall	AUC	F1
IDS-IF	Scenario_1	89%	97%	88%	94%	92%
LOF	Scenario_1	81%	93%	82%	46%	87%
IDS-IF	Scenario_2	89%	96%	88%	85%	91%
LOF	Scenario_2	81%	91%	81%	47%	87%

7.6 Conclusion

In this chapter, we proposed a novel outlier detection method with Isolation Forest (IDS-IF) to effectively detect network anomalies in SDN. The experimental results, using the well-known public network security dataset KDD, showed that IDS-IF outperforms a recent state-of-the-art outlier detection method (i.e., Local Outlier Factor (LOF) [30]) in terms of accuracy, precision, recall, AUC, F1 score, and false positives rates. This makes it a promising method to cope with the new emerging security threats in SDN.

Acknowledgments The authors would like to thank the Natural Sciences and Engineering Research Council of Canada, as well as FEDER and Grand Est Region in France, for the financial support of this research.

References

1. S. Morgan, Global ransomware damage costs predicted to reach $20 billion (USD) by 2021 [Online]. Available: https://cybersecurityventures.com/
2. L. Horwitz, The future of IoT Miniguide: The burgeoning IoT market continues [Online]. Available: https://www.cisco.com/c/en/us/solutions/internet-of-things/future-of-iot.html
3. S. Scott-Hayward, S. Natarajan, S. Sezer, A survey of security in software defined networks. IEEE Commun. Surv. Tutorials **18**(1), 623–654 (2013), First quarter 2016
4. Z.A. El Houda, L. Khoukhi, A. Hafid, ChainSecure – a scalable and proactive solution for protecting blockchain applications using SDN, in *2018 IEEE Global Communications Conference (GLOBECOM)* (2018), pp. 1–6
5. D.B. Rawat, S.R. Reddy, Software defined networking architecture, security and energy efficiency: A survey. IEEE Commun. Surv. Tutorials **19**(1), 325–346, First quarter 2017
6. Z.A. El Houda, A. Hafid, L. Khoukhi, Co-IoT: A collaborative DDoS mitigation scheme in IoT environment based on blockchain using SDN, in *2019 IEEE Global Communications Conference (GLOBECOM)* (2019), pp. 1–6
7. D. Zhou, Z. Yan, G. Liu, M. Atiquzzaman, An adaptive network data collection system in SDN. IEEE Trans. Cognit. Commun. Netw. **6**(2), 562–574 (2020)
8. Z. Abou El Houda, A.S. Hafid, L. Khoukhi, Cochain-SC: An intra- and inter-domain DDoS mitigation scheme based on blockchain using SDN and smart contract. IEEE Access **7**, 98893–98907 (2019)
9. Z.A.E. Houda, A. Hafid, L. Khoukhi, Blockchain meets AMI: Towards secure advanced metering infrastructures, in *ICC 2020 - 2020 IEEE International Conference on Communications (ICC)* (2020), pp. 1–6

10. T. Alharbi, Deployment of blockchain technology in software defined networks: A survey. IEEE Access **8**, 9146–9156 (2020)
11. Z.A.E. Houda, A. Hafid, L. Khoukhi, Blockchain-based reverse auction for V2V charging in smart grid environment, in *ICC 2021 - 2021 IEEE International Conference on Communications (ICC)* (2021), pp. 1–6
12. Y. Bengio, A. Courville, P. Vincent, Representation learning: A review and new perspectives. IEEE Trans. Pattern Anal. Mach. Intell. **35**(8), 1798–1828 (2013)
13. Y. Pei, Y. Huang, Q. Zou, X. Zhang, S. Wang, Effects of image degradation and degradation removal to CNN-based image classification. IEEE Trans. Pattern Anal. Mach. Intell. **43**(4), 1239–1253 (2021)
14. S. Mittal, M. Tatarchenko, T. Brox, Semi-supervised semantic segmentation with high- and low-level consistency. IEEE Trans. Pattern Anal. Mach. Intell. **43**(4), 1369–1379 (2021)
15. Z.A.E. Houda, A. Hafid, L. Khoukhi, BrainChain - a machine learning approach for protecting blockchain applications using SDN, in *ICC 2020 - 2020 IEEE International Conference on Communications (ICC)* (2020), pp. 1–6
16. Z. Abou El Houda, L. Khoukhi, A. Senhaji Hafid, Bringing intelligence to software defined networks: Mitigating DDoS attacks. IEEE Trans. Netw. Serv. Manag. **17**(4), 2523–2535 (2020)
17. H. Moudoud, S. Cherkaoui, L. Khoukhi, An IoT blockchain architecture using oracles and smart contracts: the use-case of a food supply chain, in *2019 IEEE 30th Annual International Symposium on Personal, Indoor and Mobile Radio Communications (PIMRC)* (2019), pp. 1–6
18. H. Moudoud, L. Khoukhi, S. Cherkaoui, Prediction and detection of FDIA and DDoS attacks in 5G enabled IoT. IEEE Netw., 1–8 (2020), vol. 35, no. 2, March/April 2021.
19. H. Moudoud, S. Cherkaoui, L. Khoukhi, Towards a scalable and trustworthy blockchain: IoT use case, in *ICC 2021 - 2021 IEEE International Conference on Communications (ICC)* (2021), pp. 1–6
20. K.P. Sinaga, M. Yang, Unsupervised k-means clustering algorithm. IEEE Access **8**, 80716–80727 (2020)
21. KDD cup 1999 dataset [Online]. Available: http://kdd.ics.uci.edu/databases/kddcup99/
22. R. Song, F. Liu, Real-time anomaly traffic monitoring based on dynamic k-NN cumulative-distance abnormal detection algorithm, in *2014 IEEE 3rd International Conference on Cloud Computing and Intelligence Systems* (2014), pp. 187–192
23. Apache Flume [Online]. Available: https://flume.apache.org/FlumeUserGuide.html
24. Apache Storm [Online]. Available: https://storm.apache.org/
25. L. Yang, H. Zhao, DDoS attack identification and defense using SDN based on machine learning method, in *2018 15th International Symposium on Pervasive Systems, Algorithms and Networks (I-SPAN)* (2018), pp. 174–178
26. M. Al-Qatf, Y. Lasheng, M. Al-Habib, K. Al-Sabahi, Deep learning approach combining sparse autoencoder with SVM for network intrusion detection. IEEE Access **6**, 52843–52856 (2018)
27. K.A. Taher, B. Mohammed Yasin Jisan, M.M. Rahman, Network intrusion detection using supervised machine learning technique with feature selection, in *2019 International Conference on Robotics, Electrical and Signal Processing Techniques (ICREST)* (2019), pp. 643–646
28. C. Yin, Y. Zhu, J. Fei, X. He, A deep learning approach for intrusion detection using recurrent neural networks. IEEE Access **5**, 21954–21961 (2017)
29. W. Wang, Y. Sheng, J. Wang, X. Zeng, X. Ye, Y. Huang, M. Zhu, HAST-IDS: Learning hierarchical spatial-temporal features using deep neural networks to improve intrusion detection. IEEE Access **6**, 1792–1806 (2018)
30. N.N. Tuan, N. Danh Nghia, P.H. Hung, D. Khac Tuyen, N.M. Hieu, N. Tai Hung, N.H. Thanh, An abnormal network traffic detection scheme using local outlier factor in SDN, in *2020 IEEE Eighth International Conference on Communications and Electronics (ICCE)* (2021), pp. 141–146
31. X. Gao, C. Shan, C. Hu, Z. Niu, Z. Liu, An adaptive ensemble machine learning model for intrusion detection. IEEE Access **7**, 82512–82521 (2019)
32. Ryu controller [Online]. Available: https://ryu.readthedocs.io/en/latest/library.html

33. Floodlight OpenFlow controller [Online]. Available: https://floodlight.atlassian.net/wiki/spaces/HOME/overview
34. Scikit-learn: Machine learning in Python. J. Mach. Learn. Res. **12**(85), 2825–2830, 2011. [Online]. Available: https://www.jmlr.org/papers/volume12/pedregosa11a/pedregosa11a.pdf
35. Google Colaboratory. [Online]. Available: https://colab.research.google.com/

Chapter 8
Deep Reinforcement Learning Modeling of a V2V Communication-Based Bike Avoidance Protocol for Increased Vehicular Flow

Sara El Hamdani, Salahedine Loudari, Mariyam Ouaissa, Mariya Ouaissa, and Nabil Benamar

8.1 Introduction

Traffic congestion is a major challenge that has serious impacts to human safety, economy, traffic efficiency, and even pollution rate. For instance, an economical study of the road congestion in Australia revealed that its cost will grow from $18.9 billion a year in 2016 to $38.8 billion a year by 2031 [1], which is very alarming.

Intelligent Transportation Systems (ITSs) is a highly active research area that aims to overcome road challenges through exploiting the technological advance achieved in the new century [1]. The enhancement in autonomous driving field opened up avenues to the application of promising Artificial Intelligence (AI) and communication technologies toward growing and improved ITSs [2].

In one side, the applications of machine learning (ML) technics [3], big data science [4, 5], and different vehicular communication technologies (i.e., EEE 802.11p [6] CNN-based 5G [7, 8]) were widely investigated throughout intensive research to enable a smart and safe driving control based on the data gathered from the traffic environment [9, 10].

In the other side, Autonomous Traffic Management (ATM) is an innovative direction of ITS that aims to completely replace the signal-based traffic control with more efficient distributed management that relies on vehicular cooperation based on Vehicle-to-Vehicle (V2V) communication [11] and hence help to decrease vehicular delay. Nonetheless, the intersection of Vulnerable Road Users (VRUs) such as cyclists and pedestrian into the full Connected and Autonomous Vehicle

S. E. Hamdani (✉) · S. Loudari · M. Ouaissa · M. Ouaissa · N. Benamar
Moulay Ismail University of Meknes, Meknes, Morocco
e-mail: s.elhamdani@edu.umi.ac.ma; s.loudari@umi.ac.ma; mariyam.ouaissa@edu.umi.ac.ma; mariya.ouaissa@edu.umi.ac.ma; n.benamar@est.umi.ac.ma

M. Ouaissa et al. (eds.), *Computational Intelligence in Recent Communication Networks*, EAI/Springer Innovations in Communication and Computing,
https://doi.org/10.1007/978-3-030-77185-0_8

(CAV) system is a remaining challenge [12] that few studies has considered [13, 14].

VRUs like cyclists share the same road with the AVs and cause driving discomfort and additional travel delay when crossing the vehicle's path. In addition, existing Advanced driver Assistance Systems (ADASs) relies on emergency braking as the main control decision to mitigate any possible vehicle-cyclist collision [15] which leads to more delays and hence to increase congestion rate.

A recently proposed protocol for Distant Bicycle Detection and Avoidance (DBDA) [14] consists on using V2V communication to broadcast an Cooperative awareness Message (CAM) in case an AV detects a bike to enable distant vehicle to have additional time to take the right action to avoid cyclist with minimum time loss. This protocol considers the necessity of making the tradeoff between cyclist protection and congestion avoidance.

Nevertheless, the major limitation of this study [14] is that the proposed DBDA protocol is a traditional programing-based algorithm. In other words, the algorithm lines are excited by the AV iteratively in the same way without any learning from past experiences. To overcome this limitation, we present a new molding of DBDA protocol based on Deep Reinforcement Learning (DRL) that provides AV with a learning process. Therefore, our proposed DRL-based DBDA enables a continuing optimization of the travel delay via repeated learning from past experiences.

A short review of related works is presented in the following section, followed by an overview of the main concepts and architectures used in our contribution. Then in our proposal, we present the DRL-based DBDA modeling. After that, a deep discussion of the effectiveness of deep learning model compared to traditional programing-based algorithm is provided. Finally, concluding remarks and future works are presented in the closing section.

8.2 Related Work

Reinforcement Learning (RL) is a dynamic programming that has been used widely in modeling different architectures aiming to solve ITS-related issues [1]. Basically, traffic environment needs to be built in order to enable AV to apply RL-based control systems. For this purpose, a microscopic simulation of highway model using OpenAI Gym framework can be integrated in various systems [2]. DRL was also largely applied for AVs control to avoid possible collision with different types of obstacles. An earlier study [16] relied on Deep Q-Learning for learning robots driving behavior in unpredictable environment.

In the same way, You et al. [3] proposed a stochastic MDP-based modeling of AV interaction in different environments that enables the vehicle to learn the driving control by imitating an expert driver style. Similarly, Emuna et al. [17] introduced a model-free model using DRL that automatically generates human policies for driving control. This study was simulated for static obstacle avoidance on highway composed with two lanes.

In addition, DRL was also [18, 19] to design distributed systems for autonomous cars control to drive in dense scenarios in which it share the road with a high flow of AVs. Furthermore, Wang and Chan [20] focused on a more challenging scenario which is on-ramp merge in which vehicles behavior is more complex and unpredictable. Accordingly, the authors proposed to apply a DRL model in order to optimize control policy. An LSTM unit was used to save experience information and hence to transport them to a deep Q-network for maximizing long-term result.

Nonetheless, all those solutions do not consider avoiding VRUs which do not belong neither to static obstacles nor to autonomously driving vehicles that are equipped with smart technologies. Few researches have considered the application of RDL for AV-pedestrian collision mitigation [21, 22] although to the best of our knowledge, no solutions were proposed for the integration of cyclists into a fully connected and autonomous system based on DRL. Thus, we propose; in this paper, a new version of DBDA protocol that relies on RDL architecture instead of traditional programming in order to enable the vehicle to safely avoid cyclists while learning from past experience and optimizing its travel delay.

8.3 Overview

8.3.1 DBDA Protocol

DBDA algorithm [14] is a distributed protocol designed to insert bicycles, as an instance of VRUs, into ATM while ensuring both cyclist's safety and traffic efficiency. Accordingly, the protocol is composed with two main processes that run successively in different AVs according to their situation.

The first part of the protocol is run when a vehicle detects a bike directly using its own sensors. Thus, the detecting vehicle V_i broadcasts an awareness message CAM to surrounding vehicles using DSRC communication technology in a range of 1 km. As a second part of the protocol, the distant vehicle V_{i+1} extracts the bike position from the received message and takes an early driving decision to mitigate the possible collision while minimizing its time loss. Accordingly, the DBDA system classifies the distance of the AV to the bike into three different zones.

Concretely as illustrated in Fig. 8.1, the accelerating zone refers to the situation where a vehicle V_{i+1} is still too far from the bike, and no immediate action is required. Contrarily, the decelerating zone refers to the case when V_{i+1} is relatively approaching to the bike, and hence, it has to decelerate and start looking for any opportunity to avoid the bike or to change the lane while minimizing any additional time loss. Finally, the avoidance zone means that the V_{i+1} is too close to the bicycle and, accordingly, must take immediate driving action to avoid the cyclist including stopping in worst case.

In this context, DBDA system enables the vehicle V_{i+1}, which is awarned by V_i, to gain additional time, and take the optimal decision alternative in terms of

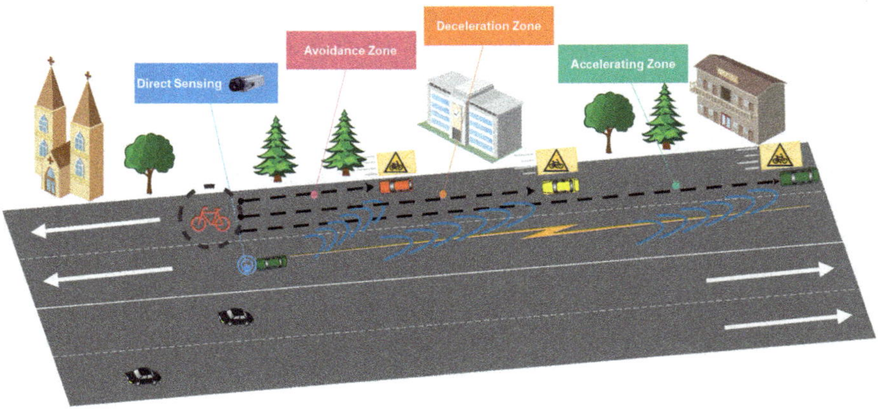

Fig. 8.1 A general illustration of a generic use case of DBDA protocol

travel delay. However, this version of the protocol is completely based on traditional programming which means that the same iterations are executed continuously by the vehicle without any learning perspectives. Thus, we propose in this paper a new modeling of DBDA protocol based on the combination of Reinforcement Learning and deep neural network technics that we review in the next subsection.

8.3.2 Deep Reinforcement Learning

ML is an area of AI that enables schemes to learn automatically from experience and progress without using an explicit programing. Thus, different ML algorithms focus on the development of computer programs that access and use data and to learn autonomously from it.

RL is one of three elementary machine learning models, along with Supervised Learning (SL) and Unsupervised Learning (UL), that is designed to enable autonomous agents to take actions in a specific environment to maximize the notion of cumulative reward [23]. Unlike SL, the RL does not require to configurate labelled pairs of input/output or to explicitly correct the suboptimal actions.

Accordingly, Markov Decision Process (MDP) is typically used to state the environment to enable a dynamic programming for RL in terms of environment states, actions, and corresponding rewards. In this context, a value-based RL process called Q-Learning can be used to optimize action-selection policy through a Q function. Therefore, a Q table can be used to discover optimal action for each state in order to maximize the Q-function.

Deep learning is part of a larger family of machine learning approaches that is based on neural networks and can be applied on difference types of ML as shown in Fig. 8.2. The DNN is inspired by the human biological system and its ability

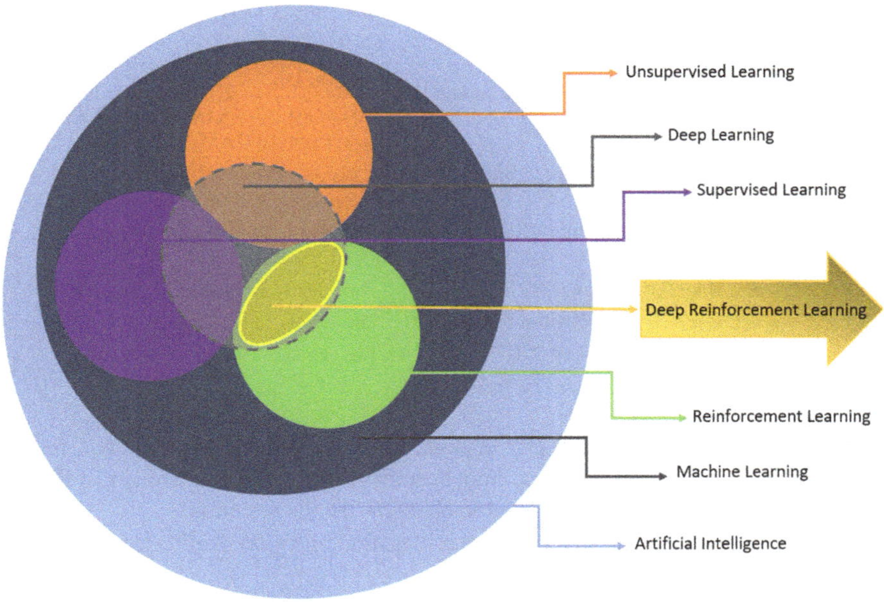

Fig. 8.2 An illustration of the position of Deep Reinforcement Learning area in the broader family of machine learning methods and artificial neural networks

to highly process information through the communication of distributed neurons. Accordingly, the Deep Q-Learning that we will use in our contribution is a DRL algorithm that relies deep learning techniques to represent the Q table and optimize the Q-function.

8.4 Markov Decision Process Modeling for DRL-Based DBDA Protocol

To model DBDA protocol as an MDP, we represent traffic situation concerning the detection of a bike gathered from sensors inputs and received awareness messages as a set of states S. As described in Table 8.1, the state S_1 represents the case when the vehicle assumes that either its sensors input or the received CAM include a bike detection directly or remotely. In turn, S_2 is the state of remote detection of a cyclist based on V2V communication. By contrast, the state S_3 represents the case when the vehicle detects a bike directly from the input of its own sensors.

The states S_4, S_5, and S_6 characterize the vehicle situations, namely, accelerating zone, decelerating zone, and avoidance zone based on its distance to the bike. In this context, we did not choose to represent those three states with only one parameter, which is the "distance," for the main reason that the vehicle can detect more than

Table 8.1 A summary of the different states representing traffic situation (the environment) in the proposed MDP for DBDA protocol

State (S_i)	Value	Description
S_1	No Bike Detecting (NBD)	No cyclist is detected via sensors input or in received CAM.
S_2	Remote Bike Detecting (RBD)	A cyclist is detected via V2V communication-based awareness.
S_3	Direct Bike Detecting (DBD)	A cyclist is detected with the vehicle's own sensors.
S_4	Far Bike Positioning (FBP)	The vehicle is the accelerating zone.
S_5	Almost Near Bike Positioning (ANBP)	The vehicle is the decelerating zone.
S_6	Close Bike Positioning (CBP)	The vehicle is the avoidance zone.

one bike either remotely or directly or both. Thus, the vehicle environment, in this case the only one distance value, will not be sufficiently representative, and different states S_4, S_5, and S_6 would be needed to characterize the situation.

The AV as an RDL agent receives a representation of the traffic environment's state annotated $S_t \in S$ at each time step.

$$t = 0, 1, 2, 3, 4, 5, \cdots, T$$

where T is the final step of the vehicle travel. According to state S_t, the AV selects a driving control action $A_t \in A$ which constitute a state-action pair annotated (S_t, A_t). Based on DBDA protocol, the control decisions founding the action set A are either to accelerate, change lane, decelerate, or stop. For each time incrementation from t to $t + 1$, the traffic state is reformed to a novel state $S_{t+1} \in S$ and take an action $A_{t+1} \in A$ for which it receives a "numerical" reward $R_{t+1} \in R$.

As shows the MDP architecture of the DRL-based DBDA illustrated in Fig. 8.3, the vehicle target is to decrease its cumulative rewards received in each time step t constituting the final return G. The procedure of getting a reward can be modeled as an arbitrary function f that links state-action pairs to received rewards. Therefore, we have A_t each time t:

$$f(S_t, A_t) = R_{t+1}$$

Depending on the DBDA algorithm, the main target of the AV is mitigating any possible collision with a cyclist by applying the optimal policy in terms of travel delay. Therefore, the main reward is the saved amount of time. Nonetheless, we chose in our model to explicitly define the reward value related to each taken action in terms of numerical points.

In this context, we set the reward R_1 of an accelerate action A_1 as plus one point "*+1 pt*" since the acceleration is the optimal decision that helps the vehicle to save time. Then we define the reward R_2 of the changing lane action A_2 as zero point

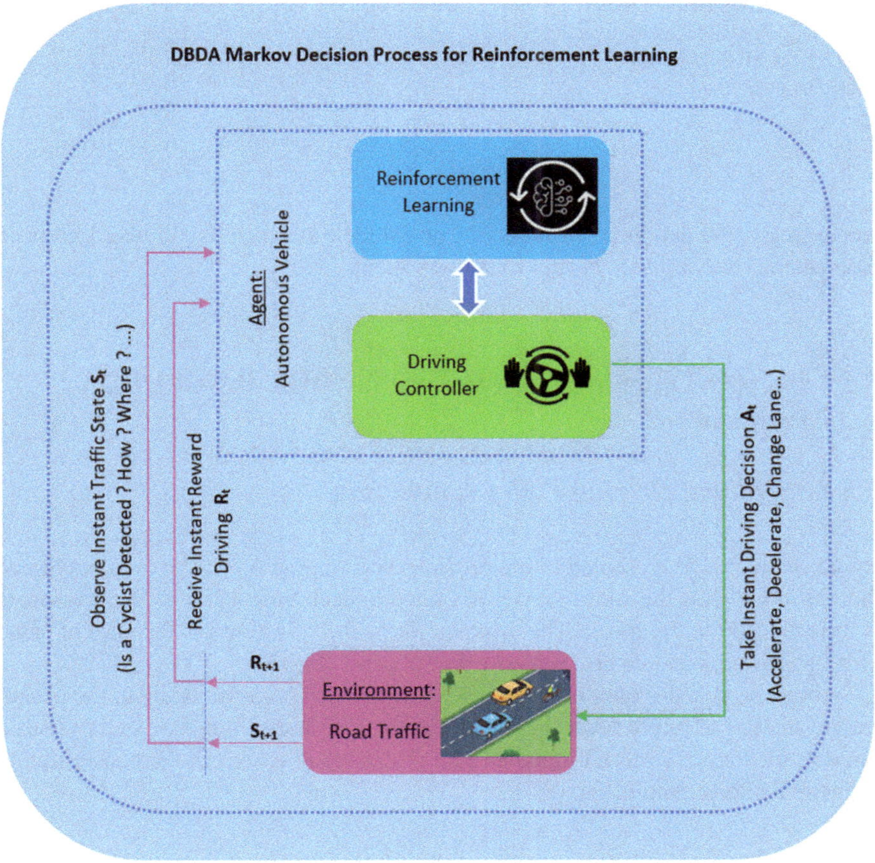

Fig. 8.3 The propose Markov Decision Process for Deep Reinforcement Learning-based modeling

"*0 pt*" because it does not help the vehicle to save or to waste time. Thereafter, we assign a negative point "*-1 pt*" as the reward R_3 related to the action A_3 which is decelerate.

Finally, we set mince ten points *"-10 pts"* as the reward R_4 for the stopping action A_4 which is the most unfavorable control decision to take for different reasons. In one side, the stop action involves braking time and starting time in addition to the stopping time. In the other side, when the vehicle takes action $A_4 = Stop$, it yields to high probability that the following vehicles take the same action.

Thus, the vehicle will not increase its own travel delay, but it will indirectly cause additional delay for next vehicles which will negatively impact the total collision rate. In this aim, we set the values of the action-reward pairs (A_i, R_i) as summarized in Table 8.2. Using action-reward, the proposed MDP model can be used to enable the vehicle to choose the optimal avoidance policy while avoiding negative rewards.

Table 8.2 A representation of the action-reward matching table for the MDP modeling of DBDA protocol

Action (A_i) value	Reward (R_i) value
A_1 = Accelerate	R_1 = 1 pt
A_2 = Change Lane	R_2 = 0 pt
A_3 = Decelerate	R_3 = -1 pt
A_4 = Stop	R_4 = -10 pts

Accordingly, the action $A_4 = Stop$ will be avoided as much as possible except in case when it is evitable to ensure road users safety.

8.5 Deep Q-Learning Architecture for DRL-Based DBDA Protocol

8.5.1 Optimal Driving Policy Estimation

Based on the MDP presented in the previous section, the AV target is to maximize the return which is the sum of rewards gained in each time step $t \in [1, T]$ where T is the final step of the travel. The discount rate $\gamma \in [0, 1]$ controls the current value of future rewards and determines how they are discounted.

In other words, discount rate γ define if the AV will focus more about the instant reward instead of future rewards. Consequently, the return G_t at time step t yields a finite result for a constant discounted rate $\gamma < 1$ based on the convergence infinite series as demonstrated in Eq. (8.1).

$$\begin{aligned} G_t &= R_{t+1} + \gamma R_{t+2} + \gamma^2 R_{t+3} + \gamma^3 R_{t+4} + \gamma^4 R_{t+5} + \cdots \\ &= R_{t+1} + \gamma \left(R_{t+2} + \gamma R_{t+3} + \gamma^2 R_{t+4} + \gamma^3 R_{t+5} + \cdots\right) \\ &= R_{t+1} + \gamma G_{t+1} \\ &= \sum_{k=0}^{\infty} \gamma^k = \frac{1}{1-\gamma}. \end{aligned} \tag{8.1}$$

To maximize its return during the travel G_t, the AV chose a policy π which is a function that relates a certain state S_t to the probabilities $\pi\ (a \mid s)$ of picking each action from that state. The value of state S_t following policy π denoted as $v_\pi(s)$ is the expected return E_π from beginning from state S at time t and applying policy π thereafter. The $v_\pi(s)$ defined with Eq. (8.2) as follows:

$$\begin{aligned} v_\pi(s) &= E_\pi\left[G_t | S_t = s\right] \\ &= E_\pi\left[\sum_{k=0}^{\infty} \gamma^k R_{t+k+1} | S_t = s\right]. \end{aligned} \tag{8.2}$$

The action value (called also Q-function) in state S_t under the policy π denoted as $q_\pi(s,a)$ represents the expected return E_π from beginning from state S_t, choosing action $A_t = a$, and following the policy π afterward. Equation (8.3) defines the action value $q_\pi(s,a)$:

$$\begin{aligned} q_\pi(s,a) &= E_\pi\left[G_t|S_t = s, A_t = a\right] \\ &= E_\pi\left[\sum_{k=0}^{\infty}\gamma^k R_{t+k+1}|S_t = s, A_t = a\right]. \end{aligned} \tag{8.3}$$

In this context, a policy π is evaluated to be the same as or better than another policy π' only if the expected return of π is superior than or equal to the expected return of π' for all states:

$$\pi \geq \pi' \text{ if and only if } v_\pi(s) \geq v_\pi{}'(s) \text{ for all } s \in S.$$

Therefore, we define the optimal state-value function denoted as $v_*(s)$ in a state St and applying the optimal policy as follows in Eq. (8.4):

$$v_*(s) = \max_\pi v^\pi(s) for\ all\ s \in S. \tag{8.4}$$

In the same way, optimal action-value function for following the optimal policy in a state S called "optimal Q-function" and denoted as q_* in Eq. (8.5) as following:

$$q_*(s,a) = \max_\pi q_\pi\ (s,a) \tag{8.5}$$

The optimal Q-function (8.5) must satisfy the Bellman Optimality Equation [24] for $Q*$ presented in Eq. (8.6) as following:

$$q_*(s,a) = E\left[R_{t+1} + \gamma \max_{a'} q_*\left(s',a'\right)\right] \tag{8.6}$$

Applying the Bellman equation [24] to DBDA protocol, the optimal $Q*$ function for each state-action pair *(s,a)* at a given time t must ensure the maximum of the expected return from beginning is the traffic state *S*, choosing the driving action *A* and applying the optimal policy afterward.

The optimal expected return in time t in this case equals the reward R_{t+1} plus the optimal anticipated discounted return that can be achieved from any probable following state-action pair *(s′,a′)*. Since the AV is following an optimal driving policy, the next state s' should be the state from which the optimal probable next action a' can be selected at the following time step $t + 1$.

Accordingly, the AV will estimate the "loss" for each policy through calculating the difference between the possible Q-function and the optimal Q-function from the Bellman Eq. (8.6) as following as follows in Eq. (8.7):

$$\begin{aligned} Loss &= q_*(s,a) - q(s,a) \\ &= E\left[R_{t+1} + \gamma\ \max_{a'} q_*\left(s',a'\right)\right] - E\left[\sum_{k=0}^{\infty}\gamma^k R_{t+k+1}\right] \end{aligned} \tag{8.7}$$

In this context, the AV will apply the loss equation to choose the optimal possible driving policy starting from a traffic state s in order to maximize the sum of the obtained rewards. Thus, it will avoid any possible collision with a bicycle while driving with the higher possible speed and avoid selecting stopping as a driving action as much as possible.

8.5.2 Deep Neural Network Architecture for Optimal Driving Learning

As we have deeply discussed in the previous subsection, the AV will analyze the traffic state in each time step based on the sensors input and the received V2V messages and calculate the Q-function for each possible driving action. Thus, the AV will be able to select the optimal driving for this traffic state that minimize the loss expressed in the aforementioned Eq. (8.7).

To provide the vehicle with a Deep Q-Learning [25], we propose in this subsection a DNN architecture for DRL-based DBDA protocol. Our proposed DNN architecture is composed with two visible layers, namely, the input layer and the output layer, in addition to the hidden layers.

The first input layer, which is a visible layer, is composed with six nodes representing the different traffic state for DBDA protocol presented previously in Table 8.1. In other words, the DNN will take the different states representing the traffic situation in each time step as an input.

The output layer, which is also the last layer, is composed with four nodes for the different possible driving action that a vehicle could apply based on DBDA protocol. Each output node presents a Q-value for one of the possible driving actions. Accordingly, the different Q-values can be compared to each other in terms of minimizing the loss value. Consequently, the AV could select the optimal action in this time step t and move to the next time step $t = 1$ which is impacted in its turn with chosen action.

The Q-values, presented by the final layer as an output, are calculated through different layers starting from the input, passing by the hidden layers, and ending in the output layer. As any typical DNN, the signals are transmitted from a layer to another based on the connections between the nodes the numerical weight of each connection is updated through the repetitive learning processing.

The DNN architecture for our DRL-based DBDA protocol is represented in Fig. 8.4. The hidden layers number varies generally in range of one layer to ten layers according to the data size and the problem complexity. Thus, we chose to use only two layers for DBDA protocol since we have a determined type of data and only few driving actions to select. For the number of hidden nodes, we set a sum of six nodes which means three nodes per hidden layer according to existing methods in this purpose [26].

Concerning the learning process, the weights of the network are initialized random as a numerical value $w \in [0, 1]$ to be updated in each learning episode. The experiences "e" are randomly memorized in replay memory D for the learning process as samples of a state s_t in time step t, the selected action a_t, the resulting

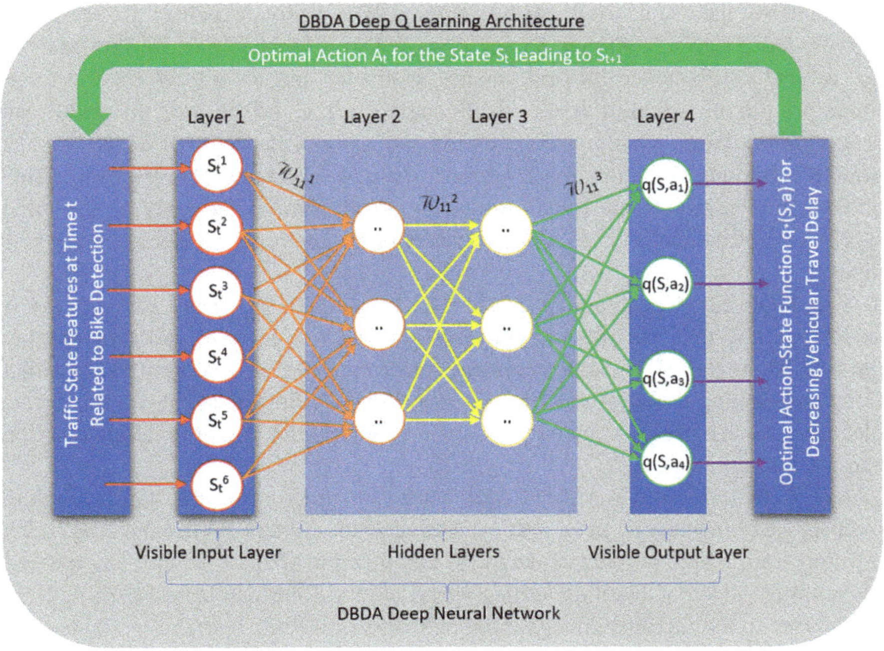

Fig. 8.4 An illustration of the Deep Q-Learning architecture for the loss optimization during the learning process

reward r_{t+1}, and the resulting state s_{t+1} as following:

$$e_t = (s_t, a_t, r_{t+1}, s_{t+1}) \tag{8.8}$$

The "Policy network," presented in Fig. 8.4, is cloned to another network called "Target network" that the Q-values resulting from first network int time t can be compared to the optimal Q-value in t + 1 in the second network. Other DNN parameters such as the learning rate and the optimizer can be set in the code and changed until finding the best learning method.

8.6 Discussion

Our contribution consists on developing a DRL-based version of DBDA protocol. In one hand, we used an MDP for bicycle avoidance process modeling that enables to represent traffic environment based on internal sensors input and external V2V awareness messages. In the other hand, we used Deep Q-Learning to enable the AV to learn how to minimize its traffic delay by taking the optimal control decision.

Therefore, the use of MDP for modeling DBDA protocol enables the vehicle to actively learn how to optimize its decisions from each experience. In this context, the vehicle can enhance its performance gradually and to outperform the traditional programming-based system as illustrated in Fig. 8.5. In fact, the traditional programming-based system maintains the same performance level despite of the available amount of data or the experiences it had. The main reason of the stable performance of traditional programming is that it consists on applying the same repetitive instructions and does not consider past experience as a part of the new input.

The MDP-based programming based on the Q-table proposed in Sect. 8.3 is an RL method that can provide conventional machine learning performance but does not allow the AV to achieve human performance level. Accordingly, by using Deep Q-Learning architectures illustrated in Fig. 8.4, the AV agent can memorize repetitive experiences, compare the Q-values, and chose the optimal state-action pair for a certain state [27].

As it is demonstrated in AI field, deep neural networks-based algorithm including Deep Q-Learning can achieve and even exceed the human performance level by learning from a high amount of data [27] compared to shallow neural network and conventional machine learning among which normal Reinforcement Learning [28]. This is due to the fact that conventional or shallow machine learning methods need

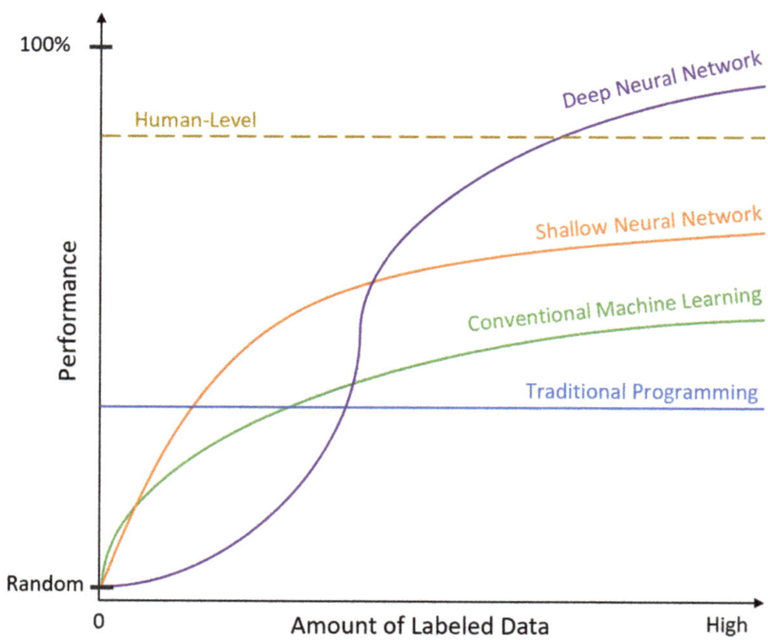

Fig. 8.5 A general comparison between different levels of machine learning models in terms of performance

a labeling of all the dataset to be able to differentiate [28]. Therefore, they attain a limited performance level since human intervention is always needed.

For all the aforementioned reasons, we argue that based on our proposed DPL-based DBDA model, the system can achieve the higher performance compared to traditional programming-based version, especially with the high amount of traffic data that the vehicle collect during its travel. In perspective work, we aim to study the optimal choice for configuration variables including the number of hidden layers and of hidden nodes and hyperparameters such as the optimizer.

8.7 Conclusion

Cyclists are VRUs that are not necessarily equipped with any sensors or communication devices like AV to be able to participate in ATM. Therefore, bike avoidance in distributed traffic management systems with the minimum time loss is a challenging task. A recent solution called DBDA protocol was proposed to cyclist avoidance using V2V communication totally based on traditional programming and does not apply any AI technics. In this paper, we have proposed a new modeling for DBDA protocol-based Deep Reinforcement Learning. Thus, we used MDP model to formalize sequential control decision-making of how the AV interact with traffic environment in different states. In addition, we applied Deep Q-Learning to enable the vehicle to learn from passed experiences and hence to be able to optimize the Q-function in terms of travel delay. Our proposed solution using Deep Reinforcement Learning provides the vehicle with a continuous development of its performance that can exceed human performance level using a high amount of data and outperform traditional programming-based DBDA that provide the vehicle with a stationary performance.

References

1. I. Bäumler, H. Kotzab, The emergence of intelligent transportation systems from a continental and technological perspective. World Rev. Intermodal. Transp. Res. **9**, 199–216 (2020). https://doi.org/10.1504/WRITR.2020.108220
2. S. El Hamdani, N. Benamar, A comprehensive study of intelligent transportation system architectures for road congestion avoidance. Lect. Notes Comput. Sci. (including Subser Lect Notes Artif Intell Lect Notes Bioinformatics) 10542 LNCS:95–106. (2017), doi:https://doi.org/10.1007/978-3-319-68179-5_9
3. A. Boukerche, J. Wang, Machine learning-based traffic prediction models for intelligent transportation systems. Comput. Netw. **181**, 107530 (2020). https://doi.org/10.1016/j.comnet.2020.107530
4. Y. Lian, G. Zhang, J. Lee, H. Huang, Review on big data applications in safety research of intelligent transportation systems and connected/automated vehicles. Accid. Anal. Prev. **146**, 105711 (2020). https://doi.org/10.1016/j.aap.2020.105711

5. S. Kaffash, A.T. Nguyen, J. Zhu, Big data algorithms and applications in intelligent transportation system: a review and bibliometric analysis. Int. J. Prod. Econ. **231**, 107868 (2021). https://doi.org/10.1016/j.ijpe.2020.107868
6. F. Arena, G. Pau, A. Severino, A review on IEEE 802.11p for intelligent transportation systems. J. Sens Actuator Networks **9**, 1–11 (2020). https://doi.org/10.3390/jsan9020022
7. D. Sirohi, N. Kumar, P.S. Rana, Convolutional neural networks for 5G-enabled intelligent transportation system: a systematic review. Comput. Commun. **153**, 459–498 (2020). https://doi.org/10.1016/j.comcom.2020.01.058
8. M. Ouaissa, M. Houmer, M. Ouaissa, An enhanced authentication protocol based group for vehicular communications over 5G networks, in *3rd Int Conf Adv Commun Technol Networking, CommNet*, (2020). doi:https://doi.org/10.1109/CommNet49926.2020.9199641
9. I. Laña, J.J. Sanchez-Medina, E.I. Vlahogianni, S.J. Del, From data to actions in intelligent transportation systems: a prescription of functional requirements for model actionability. arXiv (2020)
10. M. Ouaissa, M. Ouaissa, M. Houmer, Improved MAC design-based dynamic duty cycle for vehicular communications over M2M system, in: Springer (ed) *Data Analytics and Management*. (Singapore, 2021), pp 111–120
11. S. Hamdani El, Benamar N, Autonomous traffic management: open issues and new directions: International conference on selected topics in mobile and wireless networking (MoWNeT), pp. 1–5 (2018). https://doi.org/10.1109/MoWNet.2018.8428937
12. S. El Hamdani, N. Benamar, M. Younis, Pedestrian support in intelligent transportation systems: challenges, solutions and open issues. Transp. Res. Part C Emerg. Technol. **121**, 102856 (2020). https://doi.org/10.1016/j.trc.2020.102856
13. S. El Hamdani, N. Benamar, M. Younis, A protocol for pedestrian crossing and increased vehicular flow in smart cities. J. Intell. Transp. Syst. Technol. Planning, Oper. **0**, 1–20 (2019). https://doi.org/10.1080/15472450.2019.1683451
14. S. ElHamdani, N. Benamar, DBDA: distant bicycle detection and avoidance protocol based on V2V communication for autonomous vehicle-bicycle road share, in *2019 International Conference on Wireless Technologies, Embedded and Intelligent Systems, WITS 2019*. IEEE, (2019), pp. 1–6
15. J. Kovaceva, A. Bálint, R. Schindler, A. Schneider, Safety benefit assessment of autonomous emergency braking and steering systems for the protection of cyclists and pedestrians based on a combination of computer simulation and real-world test results. Accid. Anal. Prev. **136**, 105352 (2020). https://doi.org/10.1016/j.aap.2019.105352
16. B.Q. Huang, G.Y. Cao, M. Guo, Reinforcement learning neural network to the problem of autonomous mobile robot obstacle avoidance. Int. Conf. Mach. Learn. Cybern. ICMLC **2005**, 85–89 (2005). https://doi.org/10.1109/icmlc.2005.1526924
17. R. Emuna, A. Borowsky, A. Biess, Deep reinforcement learning for human-like driving policies in collision avoidance tasks of self-driving cars. arXiv (2020)
18. Z. Sui, Z. Pu, J. Yi, T. Xiong, Formation control with collision avoidance through deep reinforcement learning using model-guided demonstration. Proc. Int. Jt. Conf. Neural Networks, (2019) pp. 1–15. doi:https://doi.org/10.1109/IJCNN.2019.8851906
19. T. Fan, P. Long, W. Liu, J. Pan, Distributed multi-robot collision avoidance via deep reinforcement learning for navigation in complex scenarios. Int. J. Robot. Res. **39**, 856–892 (2020). https://doi.org/10.1177/0278364920916531
20. P. Wang, C.Y. Chan, Formulation of deep reinforcement learning architecture toward autonomous driving for on-ramp merge. arXiv (2017)
21. J. Li, L. Yao, X. Xu, et al., Deep reinforcement learning for pedestrian collision avoidance and human-machine cooperative driving. Inf. Sci. (Ny) **532**, 110–124 (2020). https://doi.org/10.1016/j.ins.2020.03.105
22. M. Everett, Y.F. Chen, J.P. How, Collision avoidance in pedestrian-rich environments with deep reinforcement learning. IEEE Access **9**, 10357–10377 (2021). https://doi.org/10.1109/ACCESS.2021.3050338

23. S.S. Mousavi, M. Schukat, E. Howley, Deep reinforcement learning: an overview. Lect. Notes Networks Syst. **16**, 426–440 (2018). https://doi.org/10.1007/978-3-319-56991-8_32
24. E.N. Barron, H. Ishii, The Bellman equation for minimizing the maximum cost. Nonlinear Anal. **13**, 1067–1090 (1989). https://doi.org/10.1016/0362-546X(89)90096-5
25. T. Hester, M. Vecerik, O. Pietquin, et al., Deep q-learning from demonstrations. arXiv, 3223–3230 (2017)
26. K.G. Sheela, S.N. Deepa, Selection of number of hidden neurons in neural networks in renewable energy systems. J. Sci. Ind. Res. (India) **73**, 686–688 (2014)
27. A. Araujo, B. Negrevergne, Y. Chevaleyre, J. Atif, On the Expressive Power of Deep Fully Circulant Neural Networks. arXiv (2017)
28. S. Chang, T. Cohen, B. Ostdiek, What is the machine learning? Phys. Rev. D **97**, 56009 (2018). https://doi.org/10.1103/PhysRevD.97.056009

Chapter 9
Deep Learning-Based Modeling of Pedestrian Perception and Decision-Making in Refuge Island for Autonomous Driving

Badr Ben Elallid, Sara El Hamdani, Nabil Benamar, and Nabil Mrani

9.1 Introduction

Intelligent Transportation System (ITS) is the research field that aims to solve serious traffic problematics, namely, safety and congestion issues, through communication between vehicles and the environment [1]. ITS aims to provide comfort for both drivers and passengers and to improve road safety as well as traffic fluidity and efficiency. A report of the INRIX transport data company has estimated the cost of traffic congestion in 2018 to be about $87 billion for the US economy and to be 97 hours to every American citizen wasted due to travel delay [2]. According to World Health Organization (WHO), annual road traffic deaths have reached 1.35 million in 2018 among which 21% were pedestrians [3].

To deal with road safety issues, different approaches have been proposed. Artificial Intelligence (AI) techniques such as machine learning (ML), Deep Neural Network (DNN), and Markov Decision Process (MDP) have demonstrated their ability to solve different problems and predict future behaviors and results. Such techniques have been recently extensively used in the field of ITS to overcome the most challenging transportation issues especially with the raise of autonomous driving [4]. An autonomously operating vehicle is an intelligent agent that can mimic the human driver's cognitive functions including learning and problem handling by applying AI algorithms [5].

Furthermore, smart roads do not deal only with vehicles but also with pedestrians, which need most focus and attention when building any smart road framework. Pedestrian remains the most Vulnerable Road Users (VRUs) that any malfunction

B. B. Elallid (✉) · S. E. Hamdani · N. Benamar · N. Mrani
Moulay Ismail University of Meknes, Meknes, Morocco
e-mail: badr.benelallid@edu.umi.ac.ma; s.elhamdani@edu.umi.ac.ma; n.benamar@umi.ac.ma; n.mrani@umi.ac.ma

M. Ouaissa et al. (eds.), *Computational Intelligence in Recent Communication Networks*, EAI/Springer Innovations in Communication and Computing,
https://doi.org/10.1007/978-3-030-77185-0_9

of these new technologies can put his life and health in real danger [6]. Accordingly, pedestrian perception is an active AI research area where extensive efforts are being invested to apply different machine learning methods and to enable the autonomous vehicle to recognize pedestrians and avoid any possible collision with them [7]. Thus, recent detection systems have proved their effectiveness in detecting and recognizing pedestrians among other road components using different sensors and especially in complex situations where they can be partially occluded [8].

Autonomous Traffic Management (ATM) is a new ITS research area that has a more specific objective in making the trade-off between safety and traffic efficiency by means of fully distributed and cooperative systems based on Vehicle to Vehicle (V2V) communication [6]. Thus, the new ATM architecture [9] [10] focuses on avoiding VRUs while minimizing vehicular travel delays. Accordingly, they come out with new scenarios and use cases for pedestrian detection and control decision taking that still need to be studied.

In this paper, we focus on a specific scenario in which autonomous vehicles must detect and count the pedestrians standing on a refuge island and hence apply a driving control protocol called Autonomous Pedestrian Crossing (APC) [10]. To the best of our knowledge, this specific use case was not considered by any research so far. Accordingly, we propose a new architecture that combines the Pedestrian Location Perception Network (P-LPN) and the Long Short-Term Memory (LSTM) to respectively detect/count pedestrians at a refuge island and take the convenient driving control action.

The following section presents a short survey of related works, followed by an overview of the main concepts and architectures used in our contribution. Then our proposed architecture modeling based on P-LPN and LSTM fusion for the refuge island context is presented. After that, a deep discussion of the effectiveness of our solution is given. Finally, concluding remarks and future works are presented in the closing section.

9.2 Related Work

In the last two decades, the pedestrian detection and perception field have achieved significant enhancement using effective AI techniques, namely, DNN combined with other ML algorithms [11] [12]. An earlier study demonstrated the effectiveness of the use of convolutional neural network (CNN) through reinforcement learning (RL) in terms of true positive detection. Further improvements of CNN-based pedestrian detection were proposed to train the systems on hard negatives samples. In this context, Tian et al. [13] used Task-Assistant CNN (TA-CNN) while Liu et al. [14] applied Faster Region-based CNN (FR-CNN) to further reduce the false-positive rate.

Another important challenge facing the development of high accuracy perception systems is related to the complex situations where a pedestrian is severely occluded. Wang et al. focused on heavy pedestrian occlusions detection based on Part and Context Network (PCN) and using LSTM [15].

Intra-class occlusion is even a more complex detection issue that refers to the case where a pedestrian is occluded by (an)other pedestrian(s) that makes counting or associating the right recognized head to the adequate body/parts. Therefore, Lu et al. [16] designed a solution for detecting pedestrians in a crowd scene. Accordingly, they developed a Head-Body Alignment Network (HBAN) model based on the FR-CNN framework.

Traditional pedestrian detectors including the previously reviewed system are anchor-based and hence depend on the anchor boxes of predefined scales and aspect ratios as the key element. By contrast, Liu et al. [17] proposed an anchor-free detector to save the unnecessary time and effort required for the windows configurations. In this aim, this study determined that one single center point is sufficient for pedestrian localization through a Fully Convolution Network (FCN). Consequently, the Center and Scale Prediction (CSP) was used to generate bounding boxes without the need for extra post-processing structures.

Despite all the high accuracy achieved by existing pedestrian detectors in terms of recognizing pedestrian among other components, a deeper semantic classification is required to enable AVs to understand the road scene. Therefore, Zhao et al. [18] designed a P-LPN architecture to enable AV to better percept pedestrians based on their location. In fact, pedestrian location either in the driving area or out of it can help to determine the safe pedestrians (i.e., on the sidewalk, refuge island) among the pedestrians in danger (i.e., walking on the road). Nonetheless, this research does not propose any driving decision policies to enable AVs to react based on the presented semantic perception.

A recent study [19] has proposed a more complete solution that combines the contextual road perception with the prediction of the adequate motion action to take. In this aim, the P-LPN algorithm was applied for features extraction from the input images to enable a road scene understanding. Then LSTM units were used to decide the motion decision, namely, the bake, the accelerator, and the steering wheel angle. Nevertheless, this system does not include VRUs avoidance neither for deeper road scene understanding nor for safer driving control.

In this paper, we fill this gap by providing a specific perception and control decision taking that is dedicated to AVs driving in a refuge island zone. Thus, we use a combination of P-LPN for contextual pedestrian perception on the refuge island and LSTM for the APC-based action taking [10].

9.3 Overview

9.3.1 APC Protocol

APC protocol [10] is a cooperative ATM system that aims to insert pedestrians safely in a full CAV traffic while optimizing traffic efficiency in terms of vehicular travel delay. Therefore, this protocol consists of using a pedestrian refuge island as an alternative to traditional traffic signals that require unnecessary AVs stoppings. Accordingly, pedestrians cross the road through phases and wait on the refuge after crossing each lane which enables the vehicle to be concerned by the state of the only lane on which is driving. This policy reduces the stopping or deceleration possibilities which help to decrease the time loss of the vehicle.

According to APC protocol, a blocking situation can happen if a lane gets congested for a long time due to high traffic. This can lead to a high number of waiting pedestrians queuing on the refuge island. The neighboring lane can be blocked with pedestrians that did not find any room to stand on the refuge island. Figure 9.1 illustrates a blocking situation case according to the APC protocol scenario.

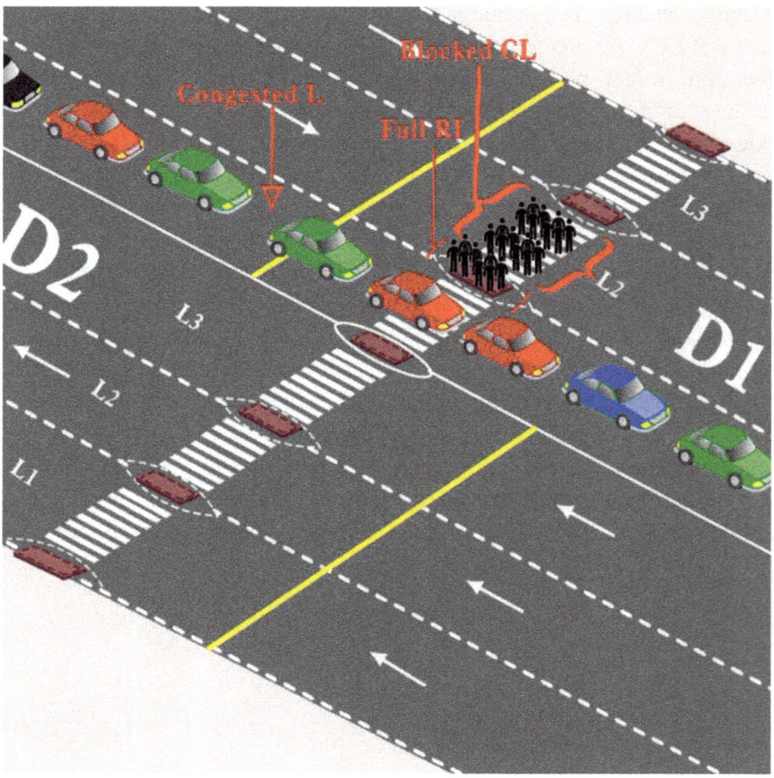

Fig. 9.1 An illustration of the blocking situation scenario according to APC protocol assumptions [10]

To deal with such situation, the APC system assumes that the AV needs to percept the state of the refuge island by counting the standing pedestrian's number and conclude if it is fully occupied and hence to stop immediately in this case. To our knowledge, no existing solution was proposed to deal with such an assumption. To fill this gap, our contribution focuses mainly on the special use case proposed in APC protocol and presents a specific perception and decision-making system based on the CNN algorithm.

9.3.2 *Deep Machine Learning*

In this section, we describe the main deep learning technologies used in AVs. We focus on CNN and LSTM, which are widely applied in AVs aspects.

9.3.2.1 Conventional Neural Network

CNN architecture has shown good results in recent years for AVs. It provides AVs with actionable information, i.e., detecting and classifying objects (lanes, traffic lights, pedestrians, crossing lines, traffic signs, etc.). Especially CNN has excellent results for image classification and object detection in real time.

CNN is based on neurons organized in layers and can learn hierarchical representations. Neurons between layers are connected by weights and biases. The first layer is the input layer that takes the data such as images or videos. Also, CNN includes a convolutional layer with multiple optimizable filters, the transform input or last hidden layers (Fig. 9.2). The purpose of the filter is to detect specific types of images feature. Thus, the depth of the convolutional layer represents the number of filters. The last layer is the output layer which predicts classification or object detection [22].

Convolution operation can be expressed as follows:

$$f_l^k(p,q) = \sum_c \sum_{x,y} i_c(x,y) . e_l^k(u,v) . \tag{9.1}$$

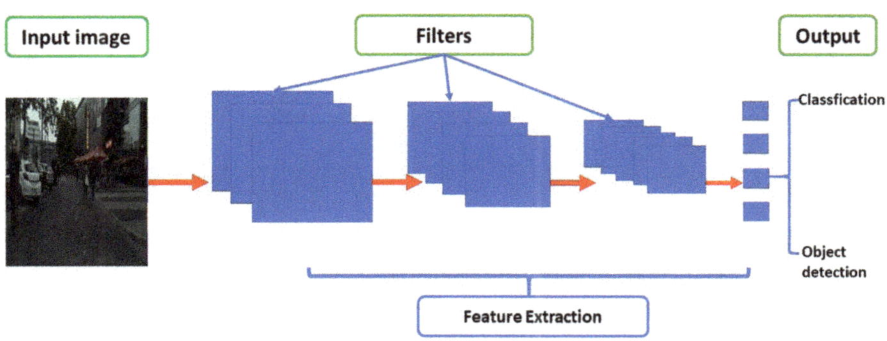

Fig. 9.2 An illustration of the Conventional Neural Network (CNN)-based architecture for pedestrian classification

where $i_c(x, y)$ is an element of input image tensor I_c which is element wise multiplied by $e_l^k\,(u, v)$ index of the kth convolutional kernel k_l of the lth layer.

Output feature map of the kth convolutional operation defines as

$$F_l^k = \left[f_l^k\,(1, 1)\,, \ldots, f_l^k\,(p, q)\,, \ldots f_l^k\,(P, Q) \right]. \tag{9.2}$$

9.3.2.2 Long Short-Term Memory

The model of Long Short-Term Memory is inherited from recurrent neural networks. An LSTM contains a unit cell that keeps state information during the time. For example, it is able to use the previous frame of the video in order to understand the current frame. The LSTM contains a memory cell, an input gate, an output gate, and a forget gate [21]. c_t and c_{t-1} represent the previous and the present cell information, respectively, as shown in Fig. 9.3. The mathematical expression is as follows:

$$f_t = \sigma \left(W_{fh} h_{t-1} + W_{fx} x_t + b_f \right), \tag{9.3}$$

$$i_t = \sigma \left(W_{ih} h_{t-1} + W_{ix} x_t + b_i \right),$$

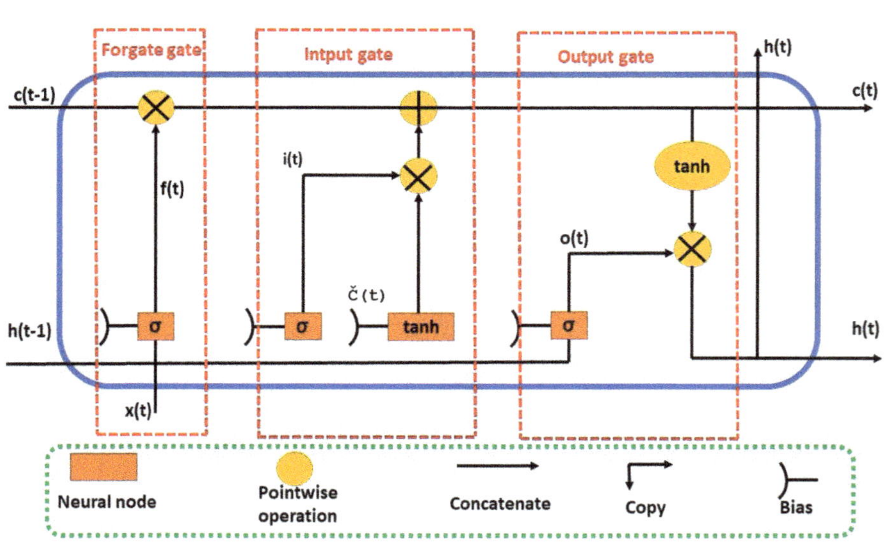

Fig. 9.3 An illustration of the general architecture of Long Short-Term Memory (LSTM) unit

$$\check{c}_t = \sigma\left(W_{\check{c}h}h_{t-1} + W_{\check{c}x}x_t + b_{\check{c}}\right),$$

$$c_t = f_t.c_{t-1} + i_t.\check{c}_t,$$

$$o_t = \sigma\left(W_{oh}h_{t-1} + W_{ox}x_t + b_o\right),$$

$$h_t = o_t.\tanh\left(c_t\right).$$

where W_i, $W_{\check{c}}$, and W_o are the weights, and the operator "." the pointwise multiplication of two vectors.

The input gate decides the new information that can be stored in the cell state when updating it, while the output gate decides what information can be output based on the cell state.

9.4 Contribution

The main goal of our approach is to enable for AV to make accurate and timely decisions in the intersection. The combination of detection and making a decision in real time is highly critical to avoid crashes in emergency situations. To this end, our method focuses on making decision depending on the number of pedestrians standing in RI.

The proposed model combines two techniques such as P-LPN and LTSM. P-LPN uses in order to detect pedestrians and compute their number in refuge island. Then we use LSTM to predict the desired action accelerator value or brake value of AV as shown in Fig. 9.4. We give more details as follows.

9.4.1 P-LPN-Based Architecture for Pedestrian Perception

P-LPN method allows detecting pedestrians in different situations such as driving scenes, sidewalks, or refuge islands [18]. As we know, most crashes happen in crosswalks or intersections, and pedestrians are Vulnerable Road Users in these situations. Thus, we focus on detecting and computing the number of pedestrians on the refuge island using P-LPN. This gives AV necessary information in order to make a decision in safe and comfortable manner.

The P-LPN consists of three modules. The first is employed to identify scenes via semantic segmentation based on Inner Cascade Network (InCNet). In addition, RPN allows to predict bounding boxes for each pedestrian in the feature maps. Finally, the fusion between the two previous modules gives the location information such as

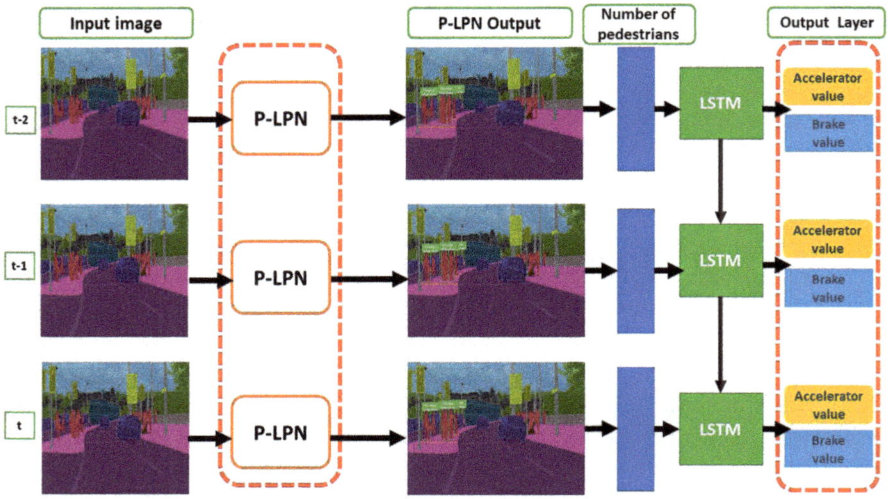

Fig. 9.4 Architecture of the proposed model with P-LPN and LSTM and using images from cityscapes dataset [20]

in driving lane, in sidewalk area, RI, etc. In our approach, we focus on the location information of pedestrians standing in RI, while the P-LPN is used to count the number of pedestrians in RI which we will use later.

The decision of AV depends on the number of pedestrians that are waiting to cross the road. Especially when pedestrians stand on refuge island, the AV cannot make a sure decision such as accelerate or a brake. For this reason, we use the P-LPN technique to detect and count the number of pedestrians. If we have a big number of pedestrians, then AV brakes in order to give pedestrians the possibility to the crossroad; otherwise, the AV accelerates.

The P-LPN schema takes consecutive frames of video and then provides location of pedestrians with their number in refuge island as shown in first part of the schema in Fig. 9.4.

9.4.2 LSTM Application for Real-Time Decision-Making

In the driving scene, time is critical especially when the AV should make a decision taking into consideration the movement of other road participants, for example, if an AV detects and counts pedestrians and the AV still far away in refuge island while the number of pedestrians changes over time. Thus, the AV will decide to accelerate because of few pedestrian numbers. However, when it comes near to the refuge island, the number of pedestrians increases to the full number the AV must brake in this way. In order to solve this problem, we propose to use LSTM architecture [21] as illustrated in Fig. 9.3 which is able to deal with sequence models and can update action in real time.

Inspiring in the model from [11], in each frame, LSTM takes a number of pedestrians computed by P-LPN, and it predicts the control command as shown Fig. 9.4. If we consider $\boldsymbol{N_t}$ to be the number of pedestrians in time t, and k represent a number of consecutive frames, the LSTM is able to update state information throughout continuous driving images.

This means the LSTM unit at the time t receives the latest memory cell c_{t-1} modulated by f_t and g_t, and the previously hidden state h_{t-1} modulated by the input gate i_t. i_t, and f_t learn to forget the previous memory or consider the current memory. This offers updates on the number of pedestrians in each selected frame in time t.

The output layer of our model is the desired control including accelerator value and brake value, represented as a vector:

$$\boldsymbol{d_t} = \left[\boldsymbol{a_t},\ \boldsymbol{b_t}\right]. \tag{9.4}$$

where $\boldsymbol{d_t}$ control command at time t, and $\boldsymbol{a_t},\ \boldsymbol{b_t}$ are accelerator and brake value, respectively.

To summarize the followed procedure as illustrated in Fig. 9.5, the model takes consecutive images from the front camera of the AV, and then P-LPN allows to

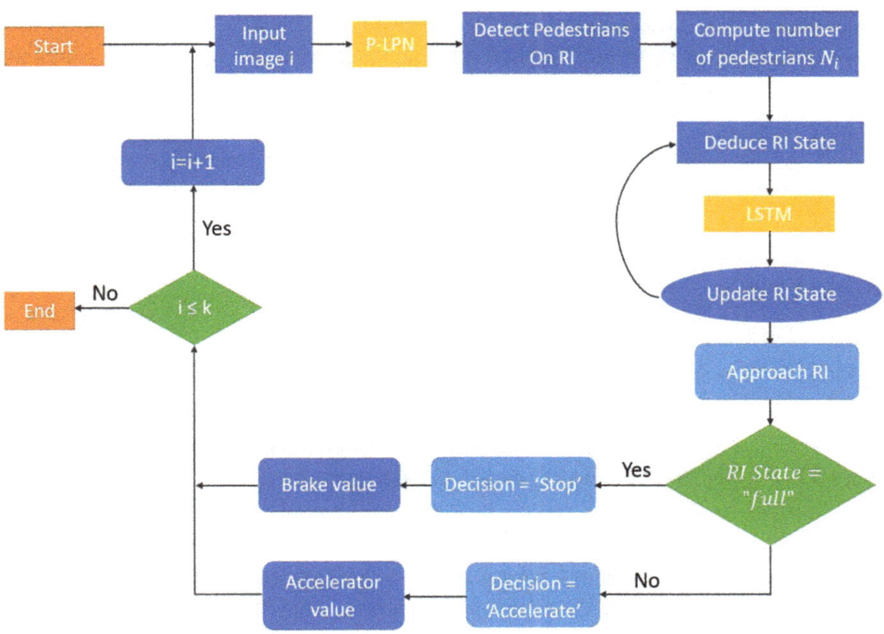

Fig. 9.5 Workflow of decision prediction procedure based on the number of pedestrians in refuge island

locate pedestrians and counts their number N_t in RI. LSTM stores N_t in order to keep or update information over time. Thus, the model tests if the RI is full or no in order to make the right decision such as stop or accelerate according to the number of pedestrians waiting to cross the street in RI.

9.5 Discussion

In this paper, we have presented a new approach in order to ensure the safety of pedestrians especially in dangerous situations like crosswalks. One of the main benefits of using our model is to reduce the travel delay of AV. It is worth noticing that other studies did not take into consideration congestion management caused by vehicles stopping. Therefore, our contribution permits the AV to stop when RI is full; otherwise, it accelerates and tries to avoid stopping. Moreover, our model permits to pedestrians to cross the road without waiting a long time. Our approach can detect and make decisions at the same time. Contrary to other techniques, their models detect only pedestrians or make decisions without taking into account the effect of pedestrians on road.

Furthermore, our method is the first to deal RI situation because the major problem happens in these cases. Especially, it is hard to identify pedestrians' intentions in intersections or crossroads with dense traffic. According to [7], the traffic light controllers are still far from being capable of ensuring pedestrian safety and comfort. For these reasons, our system is capable to reduce congestion and ensure safety for pedestrians. In addition, it facilitates AV for making decisions by detecting RI and counting the number of pedestrians and predicting the right action in a fast way.

Thus, we suggest our model in order to protect pedestrians and to a facility to AV for making the right decision. A comparison and enhancement of this paper to then existing works are provided in Table 9.1.

9.6 Conclusion

In this paper, we propose a deep learning-based model that aims to enable AV to take smart control decisions in refuge island area based on contextual perception. We used P-LPN architecture to detect pedestrians and count their number over time in order to identify the refuge state. This number is fed to LTSM to keep or change the information in real time. Therefore, an AV is able to make real-time driving decisions, such as accelerating by taking into consideration the number of pedestrians standing in this area, and hence to stop for them in the case that the refuge is fully occupied. Currently, several challenges remain in autonomous

Table 9.1 Comparison and enhancement of our model with previously published papers

Ref.	Contribution	Solution	DL techniques	Minimizing delay	Ped. detection	Driving decision
[10]	A protocol for pedestrian crossing	APC	Nontraditional programming	Yes	No	Yes
[18]	Real-time pedestrian location in complex driving scenes	P-LPN	InCNet and RPN	No	Yes	No
[19]	Multiple motion commands prediction in autonomous driving	DCN	CNN and LSTM	No	No	Yes
Our solution	DNN modeling pedestrian's detection and decision-making in RI	Perception and control in RI	P-LPN and LSTM	Yes	Yes	Yes

driving especially when dealing with the interaction between AV and pedestrians with unpredictable movements. Our perspective for future work is to study how to include AV-pedestrian interaction in our architecture in addition to perception and decision-making.

References

1. S. El Hamdani, N. Benamar, A comprehensive study of intelligent transportation system architectures for road congestion avoidance. Lect. Notes Comput. Sci. (including Subser Lect Notes Artif Intell Lect Notes Bioinformatics) **10542 LNCS**, 95–106 (2017). https://doi.org/10.1007/978-3-319-68179-5_9
2. INRIX, INRIX: congestion Costs Each American 97 hours, $1,348 A Year (2019). https://inrix.com/press-releases/scorecard-2018-us/#:~:text=Americans lost an average of, average of %241%2C348 per driver. Accessed 24 Feb 2021
3. Organization WH (2018) Global Status Report on Road Safety 2018
4. R. Abduljabbar, H. Dia, S. Liyanage, S.A. Bagloee, Applications of artificial intelligence in transport: an overview. Sustain **11** (2019). https://doi.org/10.3390/su11010189
5. J. Van Brummelen, M. O'Brien, D. Gruyer, H. Najjaran, Autonomous vehicle perception: the technology of today and tomorrow. Transp. Res. Part C Emerg. Technol. **89**, 384–406 (2018). https://doi.org/10.1016/j.trc.2018.02.012
6. S. Hamdani El, N. Benamar, Autonomous traffic management : open issues and new directions, in MoWNet'18 (ed) TAMPAS'18. IEEE, (Tangier 2018)
7. S. El Hamdani, N. Benamar, M. Younis, Pedestrian support in intelligent transportation systems: challenges, solutions and open issues. Transp. Res. Part C Emerg. Technol. **121**, 102856 (2020). https://doi.org/10.1016/j.trc.2020.102856

8. S. Aly, L. Hassan, A. Sagheer, H. Murase, Partially occluded pedestrian classification using part-based classifiers and restricted boltzmann machine model, in IEEE (ed) *16th International IEEE Conference on Intelligent Transportation Systems (ITSC 2013)*. (2016), pp. 1065–1070
9. S. ElHamdani, N. Benamar, DBDA: distant bicycle detection and avoidance protocol based on V2V communication for autonomous vehicle-bicycle road share, in *2019 International Conference on Wireless Technologies, Embedded and Intelligent Systems, WITS 2019*. IEEE, (2019), pp. 1–6
10. S. El Hamdani, N. Benamar, M. Younis, A protocol for pedestrian crossing and increased vehicular flow in smart cities. J. Intell. Transp. Syst. Technol. Planning, Oper. **24**, 1–20 (2019). https://doi.org/10.1080/15472450.2019.1683451
11. T. Gandhi, M.M. Trivedi, Pedestrian protection systems : issues, survey, and challenges. Intell. Transp. Syst. IEEE. **8**, 413–430 (2007)
12. W. Ouyang, X. Wang, Joint deep learning for pedestrian detection. (2013), https://doi.org/10.1109/ICCV.2013.257
13. Y. Tian, P. Luo, X. Wang, X. Tang, Pedestrian Detection aided by Deep Learning Semantic Tasks, (2015)
14. T. Liu, T. Stathaki, Faster R-CNn for robust pedestrian detection using semantic segmentation network. Front. Neurorobot. **12**, 1–10 (2018). https://doi.org/10.3389/fnbot.2018.00064
15. S. Wang, J. Cheng, H. Liu, et al., Pedestrian detection via body part semantic and contextual information with DNN. IEEE Trans. Multimed. **20**, 3148–3159 (2018). https://doi.org/10.1109/TMM.2018.2829602
16. R. Lu, H. Ma, Y. Wang, Semantic head enhanced pedestrian detection in a crowd. Neurocomputing **400**, 343–351 (2020). https://doi.org/10.1016/j.neucom.2020.03.037
17. W. Liu, S. Liao, W. Ren, et al., High-level semantic feature detection: a new perspective for pedestrian detection. Proc. IEEE Comput. Soc. Conf. Comput. Vis. Pattern Recognit., 5182–5191 (2019). https://doi.org/10.1109/CVPR.2019.00533
18. Y. Zhao, M. Qi, X. Li, et al., P-LPN: towards real time pedestrian location perception in complex driving scenes. IEEE Access **8**, 54730–54740 (2020). https://doi.org/10.1109/ACCESS.2020.2981821
19. X. Hu, B. Tang, L. Chen, et al., Learning a deep cascaded neural network for multiple motion commands prediction in autonomous driving. IEEE Trans. Intell. Transp. Syst., 1–12 (2020). https://doi.org/10.1109/tits.2020.3004984
20. M. Cordts, et al. The cityscapes dataset for semantic urban scene understanding, in *Proceedings of the IEEE Conference on Computer Vision and Pattern Recognition*, (2016)
21. Y. Yu et al., A review of recurrent neural networks: LSTM cells and network architectures. Neural Comput. **31**(7), 1235–1270 (2019)
22. A. Khan, A. Sohail, U. Zahoora, A.S. Qureshi, A survey of the recent architectures of deep convolutional neural networks. Artif. Intell. Rev. **53**(8), 5455–5516 (2020)

Chapter 10
Machine Learning for Hate Speech Detection in Arabic Social Media

Zakaria Boulouard, Mariya Ouaissa, and Mariyam Ouaissa

10.1 Introduction

Social media and the latest communication trends made it easier than ever to connect people from all over the globe. In January 2020, the number of active users of the Internet is 4.54 billion persons; 3.8 billion of them are on social media [1] with an increase of 9% (321 million new users) from the same time the year before. The Arab world is not spared from this global trend as the number of active users of Facebook, for example, was estimated, as of 2019, to 187 million persons in the MENA region [2]. These numbers are still expected to rise exponentially in the next few years.

With the freedom of speech, the great number of social media users is, of course, accompanied with a great number of, often conflicting, opinions. This situation can be healthy in most of the cases, but not when the expressed statements are meant to hurt or to offense others with different point of views, or even races and backgrounds.

Hate speech and cyberbullying on social media have risen to become a worrying situation in the last few years. Many of the websites and social media platforms have put on the effort to define politics to filter hate speech-related content. However, most of the solutions provided still rely heavily on reports by the users or personal monitoring by the moderators.

Z. Boulouard (✉)
Faculty of Sciences and Techniques Mohammedia, Hassan II University, Casablanca, Morocco
e-mail: zakaria.boulouard@fstm.ac.ma

M. Ouaissa · M. Ouaissa
Moulay Ismail University, Meknes, Morocco
e-mail: mariya.ouaissa@edu.umi.ac.ma; mariyam.ouaissa@edu.umi.ac.ma

M. Ouaissa et al. (eds.), *Computational Intelligence in Recent Communication Networks*, EAI/Springer Innovations in Communication and Computing,
https://doi.org/10.1007/978-3-030-77185-0_10

To overcome this situation, several researchers provided ideas based on Artificial Intelligence that can detect cyberbullying and hateful content on websites and social media [3, 4, 5]. However, most of the works available in the literature can handle content that is either in English or in the other common languages such as Spanish or French. To the best of our knowledge, the efforts that were put to filter hate speech in the Arabic content are rare.

Due to many sociopolitical reasons, Arabic social media has become a fertile area for uncontrolled hate content, and the need for hate speech detection solutions is more than necessary. In fact, these solutions can, in some cases, act as the first bastion against terrorism [6].

Hate speech detection is usually modeled as a supervised classification problem where a machine learning algorithm is trained on data that is labeled as either "hateful" or "inoffensive." Many efforts have been put to apply this process in Arabic social media, mainly Twitter, as it is a platform that is highly popular in the Middle East, especially in the Levant and Gulf areas [2]. The common point between these works is that they either focus on tweets that are in Modern Standard Arabic (MSA) [7] or using specific dialects such as the Levantine [8] or the Gulf [9] dialects.

A more interesting approach has been suggested by Alakrot et al. [10] who tried to construct a dataset based on YouTube comments on videos that are considered highly controversial in most of the Arab cultures. Such videos will surely attract hateful comments from different countries and, thus, will be in MSA as well as in different Arab dialects. We found this dataset more rich and insightful, so we used it in our approach as a base for training and testing machine learning algorithms for hate speech detection.

In the following, we will start by providing a concise, yet comprehensive, literature review on hate speech detection on Arabic social media. After that, we will present a description of the dataset provided by Alakrot et al. [10]. Then we will introduce the algorithms that we used in this study, i.e., Logistic Regression (LR), Naïve Bayes (NB), Random Forests (RF), Support Vector Machines (SVM), and Long Short-Term Memory (LSTM), as well as the metrics used to measure the performance of these algorithms, i.e., Accuracy, F1-Score, Precision, and Recall. After that, we will discuss the results and their significance.

10.2 Overview of Hate Speech Detection on Arabic Social Media

The definition of hate speech has always been source of debate for many reasons. One of them is that these definitions tend to have cultural, ethnical, and even religious influences that make them more or less biased. For this reason, it is safer to adopt definitions provided by international organizations respected around the globe. In their "Strategy and Plan of Action on Hate Speech," the United Nations

[11] have defined hate speech as "any kind of communication in speech, writing or behavior, that attacks or uses pejorative or discriminatory language with reference to a person or a group on the basis of who they are, in other words, based on their religion, ethnicity, nationality, race, color, descent, gender or other identity factor."

In recent years, with the democratization of access to social media, it has become frequent to see differences in opinion turn into hateful speech. The problem is that at first, this kind of content was not correctly supervised over the popular social media platforms, such as Facebook, Twitter, YouTube, etc., and even when they updated their politics to address hate speech, most of the efforts were heavily based on user notifications and manual interventions by administrators.

This phenomenon has caught the attention of several researchers around the globe, each focusing on content in their own languages and cultures. The greatest amount of works was, of course, applied on English-based content [5], but there were other significant works that addressed other languages, such as Spanish. For instance, Plaza-del-Arco et al. [12] compared pretrained language models for hate speech detection on Spanish tweets, while Calderon et al. [13] focused on hate tweets against migrants and refugees.

Other works suggested ideas of multilingual hate speech detection solutions that can cover multiple languages and focus on specific issues. For instance, Chiril et al. [14] focused on tweets in French and English that were considered hateful toward immigrants and women. Corazza et al. [15] were not theme specific, but their solution was able to offer hate speech detection in tweets in three languages, English, German, and Italian.

When it comes to Arabic, the number of works is surely more modest but increasing. To the best of our knowledge, the first attempts to detect hate speech in Arabic social media started in 2017 with works of Mubarak et al. [16] who classified Twitter users based how likely they would use obscene language in their tweets. Abozinadah [17] and Alakrot et al. [18] used SVM to classify tweets and YouTube comments, respectively. Alakrot et al. [18] classified their YouTube comments based on word unigrams and n-grams, while Abozinadah [17] classified his tweets based on 31 features from the user's information, social graph, etc.

The common point between these works is that they were based on social media content presented mostly in Gulf, Egyptian, and Iraqi dialects.

Other works specific to other specific dialects are available such as Mulki et al. [8] who focused on tweets in Levantine dialects. They created a dataset based on 5812 tweets labeled as "normal," "hate," or "abusive" and compared the classification performance of two algorithms, SVM and Naïve Bayes.

Applying these techniques on social media content from Maghreb countries is trickier since people in these countries are more inclined to use Arabizi writing more than Arabic. As defined by Guellil et al. [19], Arabizi is a nonstandard Romanized writing of Arabic based on a combination of Latin letters, numerals, and punctuation. Guellil et al. [19] compared several machine learning algorithms on YouTube comments related to videos about the former president "Abdelaziz Bouteflika" and his cabinet. These comments were in Algerian dialect and were written in both Arabic and Arabizi.

Other works focused on MSA (Modern Standard Arabic), which is the standard language. MSA is mostly used for formal communication purposes, and the words are usually composed of a root fitted to a stem template, prefixes, and suffixes that may act as pronouns or conjunctions. Mubarak et al. [16] have provided a rich dataset of 32 K comments on news articles from Aljazeera's website. Most of these comments were in MSA, with some exceptions in different Arabic dialects. These comments were classified as either "obscene," "offensive," or "clean." The percentage of "offensive" comments was the highest with 79% against 19% "clean" comments and 2% "obscene" comments.

Other attempts tried to combine Arabic with other languages. According to Guellil et al. [19], a great number of Arabs are bilingual. In the Middle East region, the second most common language after Arabic is English, while in the Maghreb region, it is French. In their work, Ousidhoum et al. [20] tried to combine the three languages in a multilingual multi-aspect hate speech detection process. They created a dataset of 13 K tweets in the three languages annotated for five different aspects, and then they compared the performances of Logistic Regression (LR), Long Short-Term Memory (LSTM), and Bidirectional Long Short-Term Memory (Bi-LSTM). The performances are measured using the Micro-F1 and Macro-F1 scores of different classification tasks: Single Task Single Model (STSM), Single Task Multilingual Model (STML), and Multitask Multilingual Model (MTML).

10.3 Natural Language Processing

Data mining is a set of techniques that enable us to extract meaningful information from masses of data.

When the data consists of unstructured full texts expressed in "Natural" human languages, proper preprocessing is needed in order to extract statistical features that can be fed to machine learning algorithms.

This procedure, also known as "Natural Language Processing," or shortly as NLP, consists of a number of steps going from tokenization to the removal of stop words and text vectorization.

The idea behind tokenization is to divide a text into a list of words, known as tokens divided based on a separator (usually the "space" character).

For example, let's consider we have a sentence (*S*) that says, "Hate speech is affecting both the user and the consumer of content."

Tokenizing this sentence would result in

$$S = \Big['Hate', 'Speech', 'is', 'affecting', 'both', 'the', 'user',$$

$$'and', 'the', 'consumer', 'of', 'content' \Big].$$

This sentence can be further distilled by removing "stop words." The stop words are usually pronouns, prepositions, articles, and conjunctions that don't carry significant meaning on their own.

Another step in the distillation process is stemming, which will be explained in the next subsection. In short, stemming is the removal of extra prefixes and suffixes and the extraction of the word's root.

After applying stemming and removing stop words from the sentence (*S*), the result would become

$$S = \left['Hate', 'Speech', 'affect', 'user', 'consumer', 'content'\right]$$

In order to feed this distilled text into machine learning algorithms for further analysis, it needs to be presented as a vector of statistical features. For instance, if we replace each of these words by the number of times it appeared from the beginning of this chapter until this particular line, the sentence (*S*) would be represented by a vector (*V*) that looks like

$$V = [21, 23, 1, 3, 1, 11]$$

Further preprocessing can be done by using different NLP techniques such as stemming, lemmatization, and the Bag of Words, while other statistical features can be extracted with techniques such as "Term Frequency (TF)" and "Term Frequency-Inverse Document Frequency (TF-IDF)."

10.3.1 Stemming

Stemming is the process of removing prefixes and suffixes in order to bring the word to its root. This root can either be the dictionary root or not. When we bring a word to its dictionary root, also known as lemma, this process is called "lemmatization." Stemming, in general, can present a few complications; the most famous of them are "overstemming" and "understemming."

Overstemming is when a larger part of the word than usual is removed, and the resulting stem is not the desired one. For instance, when we try to stem words such as "university" and "universal," the resulting root will be "univers" in both cases, which is not correct.

Understemming, on the other hand, can cause two similar words to fall into different stems. For example, the stemming of "data" and its singular "datum" can fall into "dat" and "datu," respectively, which is, again, not correct.

10.3.2 Bag of Words and Term Frequency

The Bag of Words (or BoW) is a method of text representation where the text is decomposed into a set of fully independent words disregarding any semantic or grammatical relationships whatsoever.

It is commonly used for classification tasks where the Term Frequency (or TF) of each word can be considered as a feature when training a classification algorithm.

Term Frequency is an approach that is similar to the BoW, but the difference is that it represents how frequent a word is in a document. It is calculated by dividing the number of occurrences of a keyword n_i divided by the total number of the keywords in the document N as in the following equation:

$$TF = \frac{n_i}{N} \tag{10.1}$$

However, TF is not usually the best feature to extract since stop words such as "the," "in," etc., tend to have the highest frequency in general. So having the highest frequency doesn't necessarily mean that a word is relevant.

To solve this issue, another approach is suggested, which is "Term Frequency-Inverse Document Frequency" (or TF-IDF).

10.3.3 Term Frequency-Inverse Document Frequency (TF-IDF)

TF-IDF is an indicator that reflects the importance of a word in a document, collection, or corpus. Different from the TF approach, TF-IDF can give more weight to words that appear frequently in a set of documents than those that are known to be frequent in general (i.e., stop words).

The TF part is the Term Frequency calculated using Eq. (10.1), while the IDF part is described as an indicator that shows how rare a word is across all the documents of a corpus. The *IDF* is inversely proportional to the number of occurrences. So if the *IDF* is closer to 0, this means that the word is very common.

The *IDF* is expressed using the following equation:

$$IDF = \log\left(\frac{N_D}{N_i}\right) \tag{10.2}$$

where N_D is the number of documents in the corpus, while N_i is the number of documents that contain the keyword under study.

TF-IDF being the product of (10.1) and (10.2), its equation will become

$$TF - IDF = \frac{n_i}{N}\log\left(\frac{N_D}{N_i}\right) \tag{10.3}$$

10.4 Dataset

10.4.1 YouTube

By the start of 2021, Internet Live Stats [21] has reported that the number of videos being viewed on YouTube is over 88,416 per second. These videos are uploaded at a rate of 500+ hours of content per minute and are viewed by 2 billion monthly logged-in users from more than 100 countries and speaking over 80 different languages [22].

However, compared to other social media platforms, YouTube's supervision on the comments section, below the videos, is minimalistic. This leaves more room for the publication of offensive content, and whenever some content is reported as such, it takes time to remove it.

It is also the case in the MENA region where the Arabic content is highly exposed to offensive comments, especially when the concerned videos provide content this is regarded as "controversial" in the oriental culture. This is also the case when the content is subject to opposing opinions, where the comments tend to focus more on insulting the opposing side rather than having a healthy exchange of opinions.

When publishing content, some users tend to publish specific content that offers higher risks of offensive comments, just to have "more interaction" in their YouTube channels. This creates a suitable environment for insults, hate speech, and social stigmatization. Table 10.1 illustrates some examples of hateful comments on Arabic videos on YouTube. These examples are from the dataset provided by Alakrot et al. [10].

Table 10.1 Hateful comments from YouTube videos collected by Alakrot et al. [10] and their translations to English

Translation to English	YouTube comments in Arabic	Video topic
What is this shit? You are not Ahlam (dreams). You are nightmares. You look just like Michael Jackson the day before he dies.	ماهد الخراااء انت لست احلام إنتي كوابيس تشبهين مايكل جاكسون قبل وفاته ب بيوم	Controversial singer
The idiot, she has psychological issues for sure. She's the silliest and most stupid person I have ever seen in my life (her voice, her looks, and her actions).	الهبلة هي اكيد مريضة نفسيا اتفه و ابيخ واسخف مخلوقة بشوفها بحياتي "صوتا وشكلا وتصرفاتا"	Controversial singer
I just want to know why the transsexuals forget to transform their voices, looool.	انا عايزة اعرف حاجة ...ليه المتحول الجنسي بينسى يحول صوته كمان ....هههههههههههههههه	Transsexual coming out
This is the first time I see such a freak show.... Who is like me?	اول مرة اشوف هالمسخ ... مين مثلي؟؟	Transsexual coming out

10.4.2 Alakrot's YouTube Comments Collection [10]

For their dataset, Alakrot et al. [10] collected 15,050 comments from YouTube channels that tend to publish controversial videos about celebrities and people who are known in the Arabic social media. This type of videos can have a direct impact on YouTube viewers and incite them to comment using offensive language.

Out of a collection of 150 videos, Alakrot et al. [10] selected nine videos that have the highest number of comments to build their dataset, under the assumption that the higher the number of comments is, the most likely it will have a greater number of hateful comments.

A study on the diversity of dialects used in the comments from these videos has shown that the top 3 dialects present are Iraqi, Gulf, and Egyptian. Three annotators were selected from three different Arab countries. The first two came from highly represented countries, Iraq and Egypt, while the third is from Libya, a country that is not very represented in the dataset. The annotators were asked to label the comments as either "offensive" or "inoffensive." The three of them agreed on the same labeling (offensive or inoffensive) of 10,715 comments (71% of the dataset), while two out of three agreed on 5817 (39% of dataset) being offensive.

Based on the annotations on the comments, Alakrot et al. [10] had to make a final decision on which comments are considered hateful and which are not. For that, they adopted two scenarios: The first scenario is to label offensive the comments on which all three annotators agree and label the rest as inoffensive. The second scenario is to label offensive the comment on which at least two annotators agree and labeling the rest as inoffensive.

Our study will be based on comments labeled using the second scenario as we found that there were more errors in the labeling that is based on the first scenario. Table 10.2 displays some YouTube comments along with the annotators' labels and the final labeling based on the second scenario.

- "P" is for positive (offensive) comments.
- "N" is for negative (inoffensive) comments.

As a reminder, Alakrot et al. [10] made their dataset available to public use in the following link:[1] https://goo.gl/27EVbU.

[1] [Accessed: 28-Feb-2021]

Table 10.2 Hateful comments from YouTube videos collected by Alakrot et al. [10] along with their annotations and final labeling based on the second scenario

Final labeling	Annotator 3	Annotator 2	Annotator 1	YouTube comments
P	N	P	P	شمس مثقفة و جميلة وايد ... و احلام هههههععع ما اقول
N	N	N	P	احلام اني حبيتها اكثر لان انظلمت كثير من الناس هي امراه قويه وكبيره في عيونا وليسه محتشم
N	N	N	N	أحلام فنانه طيبه علي نيتها وبتساعد الناس وانا بحبها علشان علي طبيعيتها
P	P	P	P	اتفوو على شراميط مين خلف هالوسخات ماعندهم اهل يحكمهم

10.5 Methodology

10.5.1 Preprocessing

According to Aljarah et al. [23], data preprocessing includes several steps starting from collecting, annotation, cleaning, normalization, tokenization, and, finally, vectorization.

For their dataset, Alakrot et al. [10] organized the extracted YouTube comments in an Excel file with twenty columns (index, user, comment text, annotations, label, etc.). The columns that interest us most are the ones that contain the comment text and the final labeling based on the second scenario as mentioned earlier. That being said, the collecting and annotation steps were already done.

The first step for us was the cleaning. We started by filtering the lines with NULL values, reducing the number of comments in this file to around 11 K comments, 4748 of them were considered hateful, while 6250 were considered inoffensive. After removing the NULL lines, we removed the punctuations and diacritics as they tend to be used a lot in Arabic. After that, we used the Natural Language Toolkit (NLTK) [24] to go through the comments to remove the stop words.

The next step is normalization. In social media, it is very common to use elongations, which are excessive uses of some particular letters to express some feelings such as excitement. For instance, it is very common in English-speaking social media to see comments like "**soooo cuuuuuute**," which can be normalized to "**so cute**." The situation is similar in Arabic social media, where we can see elongations of particular letters. For example, it is common to see comments like "'**أحبيييييك**'" which can be normalized to "'**أحبك**'." Another case where normalization is necessary is for the "Alef" letter (ا) that can take many forms according to the position and shape of the "Hamza" character (ء). So we normalized the variants

(أ آ ا إ) by replacing them with the same (ا). Another special case for normalization is the use of (گ), which represents a tentative to introduce the Latin letter (**G**) to Arabic. This letter is very common in Arabic dialects though it doesn't exist in the classical Arabic or the Modern Standard Arabic (MSA). As part of the normalization process, we replaced the (گ) letter by its closest relative in MSA, the letter (ك), which is the Arabic equivalent of the Latin letter (**K**).

For stemming and tokenization, we used the NLTK library.

For the vectorization, we chose to adopt the TF-IDF approach that was described in Sect. 10.3 as it was the most relevant to our use case.

After preprocessing, the data was split into two subsets:

- Training dataset, which contains 80% of the data.
- Testing dataset, which contains the remaining 20%.

10.5.2 Algorithms

For this study, we chose to compare five machine learning algorithms:

10.5.2.1 Logistic Regression (LR)

Logistic Regression (LR) is defined by Nayebi [25] as an algorithm that can display the impact that a number of independent variables can have on the probability of occurrence of a dependent variable.

As for the linear regression, the Logistic Regression examines assumed relationships between a dependent variable and a set of independent variables. The main difference between these two models is that in the linear regression, the dependent variable is quantitative, while in the Logistic Regression, the dependent variable is a binary (dichotomous or binomial) variable. Mathematically, a Logistic Regression estimates a multiple linear regression described as

$$\ln \frac{P(1 \mid X)}{1 - P(1 \mid X)} = \beta_0 + \beta_1 X_1 + \beta_2 X_2 + \cdots + \beta_n X_n$$

$\beta_i \, for \; i \in \{1, 2, \ldots, n\}$ being the regression coefficients
$X_i \, for \; i \in \{1, 2, \ldots, n\}$ being the independent variables
β_0 being the intercept

This equation is a regression because it displays the relationship between a dependent variable and a set of independent variables. It is also logistic because the probability of occurrence of a class can be expressed as a logistic distribution function

$$P(1 \mid X) = \frac{e^{\beta_0+\beta_1 X_1+\beta_2 X_2+\cdots+\beta_n X_n}}{1+e^{\beta_0+\beta_1 X_1+\beta_2 X_2+\cdots+\beta_n X_n}}$$

Naïve Bayes (NB)

Naïve Bayes (NB) [26] is an algorithm used mainly for classification purposes. The idea behind this algorithm is to provide classifications based on probabilities of events. This classification technique is based on a combination of the Bayes rule and the independence between variables. Though the independence condition is neglected in general, this algorithm still provides good results.

The Bayes rule is described by the following equation:

$$P(A \mid B) = \frac{P(B \mid A) * P(A)}{P(B)}$$

Compared to other classification algorithms, Naïve Bayes can perform very well since it needs less time for computing the event probabilities and for training. The only downfall is that it considers that the variables are fully independent, which is not the case in general.

10.5.2.2 Random Forests (RF)

Random Forests (RF) [27] is an ensemble learning algorithm used for classification purposes.

It starts by training each decision tree of the forest on a subset of the training data. The data affected to each tree is selected randomly using the "Bagging" process. This allows the RF algorithm to provide better performance by decreasing the variance of the model without increasing the bias.

The Random Forests algorithm will, later, combine the results of the decision trees for its final outcome. In the case of binary classifications, the return of the Random Forest algorithm is, usually, the class that was predicted by most of the decision trees.

10.5.2.3 Support Vector Machines (SVM)

Support Vector Machines (SVM) [28] is a classification algorithm that tries to find an optimal hyperplane in an N-dimensional space (N being the number of features) that can classify the data into categories. The best classification is where data points are the furthest possible to the hyperplane outliers.

The hyperplanes can be seen as boundaries that help separate the data to the desired classes. The dimensions of a hyperplane depend on the number of features. For example, if the number of input features is two, the hyperplane will be a line, while for three features, the hyperplane will become a two-dimensional plane.

10.5.2.4 Long Short-Term Memory (LSTM)

Long Short-Term Memory (LSTM) [29] is a recurrent artificial neural network architecture that can provide feedback connections, unlike the standard feed forward architectures. The scope of applications of LSTM covers different fields such as handwriting recognition, speech and video recognition, as well as NLP-related tasks. LSTM networks can be versatile as they can solve classification, processing, as well as prediction problems.

Compared to normal recurrent neural networks, LSTMs can conserve a more stable error level that can be back-propagated over time and layers. This will allow the network to keep learning over many epochs.

10.5.3 Evaluation Metrics

The performance of the abovementioned algorithms will be measured using four evaluation metrics:

- Accuracy
- F1-Score
- Precision
- Recall

For binary classification, these metrics depend on four variables:

- True Positives (TP): These are the correctly predicted positive values (in our case, these will be comments that were predicted correctly as "offensive").
- True Negatives (TN): These are the correctly predicted negative values (in our case, these will be comments that were predicted correctly as "inoffensive").
- False Positives (FP): These are the wrongly predicted positive values where they should actually be negatives (in our case, these will be comments that were predicted as "offensive" while they should actually be "inoffensive").
- False Negatives (FN): These are the wrongly predicted negative values where they should actually be positive (in our case, these will be comments that were predicted as "inoffensive" while they should actually be "offensive").

Accuracy is the ratio of the correctly predicted values (True Positives and True Negatives) over the total number of values. This metric is usually efficient when the values of the False Negatives and False Positives are almost the same.

Accuracy can be expressed using the following equation:

$$Accuracy = \frac{TP + TN}{TP + TN + FP + FN}$$

Precision is the ratio of the correctly predicted values (True Positives) over the total number of positive predictions (True Positives and True Negatives).

Precision can be expressed using the following equation:

$$Precision = \frac{TP}{TP + TN}$$

Recall, also called "Sensitivity," is the ratio of the correctly predicted values (True Positives) over the total number of values related to the class in question.

Recall can be expressed using the following equation:

$$Recall = \frac{TP}{TP + FN}$$

F1-Score is the weighted average between Precision and Recall. Compared to Accuracy, the F1-Score can be more useful in case class distribution is not even.

F1-Score can be expressed using the following equation:

$$F1 - Score = \frac{2 * Recall * Precision}{Recall + Precision}$$

10.6 Experimental Results

The algorithms were fully developed in Python 3 using the "Scikit-Learn" library[2] for the Logistic Regression, the Random Forests, and the Support Vector Machines. For the Long Short-Term Memory algorithm, we opted for the "TensorFlow 2.0" library.[3] The development was done using the "Colaboratory" cloud platform.[4]

Table 10.3 represents the classification performance of the abovementioned algorithms (LR, NB, RF, SVM, and LSTM) measured by the four adopted metrics (Accuracy, Precision, Recall, and F1-Score). As we mentioned earlier, the feature on which we based our study is the TF-IDF (Term Frequency-Inverse Document Frequency).

Table 10.3 Classification performance scores

Algorithm	Accuracy	Precision	Recall	F1-Score
LR	0.8	0.8	0.69	0.74
NB	0.78	0.86	0.55	0.67
RF	0.76	0.9	0.47	0.62
SVM	0.81	0.83	0.71	0.76
LSTM	0.82	0.92	0.74	0.82

[2]https://scikit-learn.org/stable/ [Accessed: 28-Feb-2021]

[3]https://www.tensorflow.org/ [Accessed: 28-Feb-2021]

[4]https://colab.research.google.com/ [Accessed: 28-Feb-2021]

All scores considered LSTM provided better results, followed by SVM. The scores between the other algorithms vary according to the metric.

For Accuracy, the best result was provided by LSTM with 0.82, closely followed by SVM with 0.81, and then LR with 0.8. NB and RF provided the worst Accuracy performance with 0.78 and 0.76, respectively.

For Precision, the best result was provided by LSTM with 0.92, closely followed by RF with 0.9, and then NB with 0.86. SVM and LR provided the worst Precision performance with 0.83 and 0.8, respectively.

For Recall, the order was similar as in Accuracy as the best result was provided by LSTM with 0.74, followed by SVM with 0.71, and then LR with 0.69. NB and RF provided the worst Recall performance with 0.55 and 0.47, respectively.

For F1-Score, the best result was provided by LSTM with 0.82, followed by SVM with 0.76, and then LR with 0.74. NB and RF provided the worst F1-Score performance with 0.67 and 0.62, respectively.

10.7 Conclusion and Perspectives

In this study, we compared the classification performance of five algorithms for hate speech detection purposes. The algorithms tested were Logistic Regression (LR), Naïve Bayes (NB), Random Forests (RF), Support Vector Machines (SVM), and Long Short-Term Memory (LSTM). The performance metrics used were Accuracy, Prediction, Recall, and F1-Score.

The dataset we used was provided by Alakrot et al. [10]. This openly available dataset contains YouTube comments from videos with content that is considered highly controversial in most of the Arab cultures. These comments were labeled as either "P" for positive, offensive comments, and "N" for negative, inoffensive comments.

We preprocessed the dataset to extract the TF-IDF before feeding it to the classification algorithms. LSTM provided the best performance all scores considered, followed by SVM. The performances of the other algorithms vary according to each score.

In future works, more fine-tuning will be needed in the preprocessing step (enriching the stop words list according to the different dialects, the use of different stemmers, etc.). We can also think about extracting more features such as BoW, TF, as well as contextual features.

References

1. S. Kemp, "Digital 2020: 3.8 billion people use social media – We Are Social," (2020). [Online]. Available: https://wearesocial.com/blog/2020/01/digital-2020-3-8-billion-people-use-social-media. Accessed 21 Feb 2021
2. D. Radcliffe, H. Abuhmaid, Social Media in the Middle East: 2019 in Review, SSRN Electronic J., (2020)

3. S. Modha, P. Majumder, T. Mandl, C. Mandalia, Detecting and visualizing hate speech in social media: a cyber watchdog for surveillance. Expert Syst. Appl. **161**, 113725 (2020)
4. P. Kapil, A. Ekbal, A deep neural network based multi-task learning approach to hate speech detection. Knowl.-Based Syst. **210**, 106458 (2020)
5. F.E. Ayo, O. Folorunso, F.T. Ibharalu, I.A. Osinuga, Machine learning techniques for hate speech classification of twitter data: state-of-the-art, future challenges and research directions. Comput. Sci. Rev. **38**, 100311 (2020)
6. W. Alhalabi et al., Social mining for terroristic behavior detection through Arabic tweets characterization. Futur. Gener. Comput. Syst. (2020)
7. H. Mubarak, A. Rashed, K. Darwish, Y. Samih, A. Abdelali, Arabic offensive language on twitter: analysis and experiments. arXiv (2020)
8. H. Mulki, H. Haddad, C. Bechikh Ali, H. Alshabani, L-HSAB: a Levantine Twitter dataset for hate speech and abusive language, in *Proceedings of the Third Workshop on Abusive Language Online*, (2019), pp. 111–118
9. R. Alshalan, H. Al-Khalifa, A deep learning approach for automatic hate speech detection in the saudi twittersphere. Appl. Sci. (Switzerland) **10**(23), 1–16 (2020)
10. A. Alakrot, L. Murray, N.S. Nikolov, Dataset construction for the detection of anti-social behaviour in online communication in Arabic. Procedia Comput. Sci. **142**, 174–181 (2018)
11. United Nations, United Nations Strategy and Plan of Action on Hate Speech, (2019)
12. F.M. Plaza-del-Arco, M.D. Molina-González, L.A. Ureña-López, M.T. Martín-Valdivia, Comparing pre-trained language models for Spanish hate speech detection. Expert Syst. Appl. **166**, no. March 2020, 114120 (2021)
13. C. Arcila Calderón, D. Blanco-Herrero, M.B. Valdez Apolo, Rechazo y discurso de odio en twitter: análisis de contenido de los tuits sobre migrantes y refugiados en español/rejection and hate speech in twitter: content analysis of tweets about migrants and refugees in Spanish. Revista Española de Investigaciones Sociológicas **172**, 21–39 (2020)
14. P. Chiril, F. Benamara Zitoune, V. Moriceau, M. Coulomb-Gully, A. Kumar, Multilingual and Multitarget Hate Speech Detection in Tweets, *Actes de la Conférence sur le Traitement Automatique des Langues Naturelles (TALN) PFIA 2019. Volume II : Articles courts*, **4**, (2019), pp. 351–360
15. M. Corazza, S. Menini, E. Cabrio, S. Tonelli, S. Villata, A multilingual evaluation for online hate speech detection. ACM Trans. Internet Technol. **20**(2), 1–22 (2020)
16. H. Mubarak, K. Darwish, and W. Magdy, Abusive Language Detection on Arabic Social Media, in *Proceedings of the First Workshop on Abusive Language Online*, (2017), pp. 52–56
17. E. Abozinadah, *Detecting Abusive Arabic Language Twitter Accounts Using a Multidimensional Analysis Model* (George Mason University, 2017)
18. A. Alakrot, L. Murray, N.S. Nikolov, Towards accurate detection of offensive language in online communication in Arabic. Procedia Comput. Sci. **142**, 315–320 (2018)
19. I. Guellil, A. Adeel, F. Azouaou, S. Chennoufi, H. Maafi, T. Hamitouche, Detecting hate speech against politicians in Arabic community on social media. Int. J. Web Inf. Syst. **16**(3), 295–313 (2020)
20. N. Ousidhoum, Z. Lin, H. Zhang, Y. Song, D.-Y. Yeung, Multilingual and Multi-Aspect Hate Speech Analysis, in *Proceedings of the 2019 Conference on Empirical Methods in Natural Language Processing and the 9th International Joint Conference on Natural Language Processing (EMNLP-IJCNLP)*, (2019), pp. 4674–4683
21. Internet Live Stats, 1 Second – Internet Live Stats, (2021). [Online]. Available: https://www.internetlivestats.com/one-second/#youtube-band. Accessed 28 Feb 2021
22. YouTube Blog, "YouTube for Press," (2021). [Online]. Available: https://blog.youtube/press/. Accessed 28 Feb 2021
23. I. Aljarah et al., Intelligent detection of hate speech in Arabic social network: a machine learning approach. J. Inf. Sci., 016555152091765 (2020)
24. NLTK, Natural Language Toolkit — NLTK 3.5 documentation, (2021). [Online]. Available: https://www.nltk.org/. Accessed 02 Mar 2021

25. H. Nayebi, Logistic regression analysis, in *Advanced Statistics for Testing Assumed Casual Relationships*, (Springer, Cham, 2020), pp. 79–109
26. G. I. Webb, E. Keogh, R. Miikkulainen, R. Miikkulainen, M. Sebag, Naïve Bayes, in *Encyclopedia of Machine Learning*, (Springer US, 2011), pp. 713–714
27. Y. Liu, Y. Wang, J. Zhang, New machine learning algorithm: Random forest, in *Lecture Notes in Computer Science (including subseries Lecture Notes in Artificial Intelligence and Lecture Notes in Bioinformatics)*, (2012), vol. 7473 LNCS, pp. 246–252
28. S.H.H. Mehne, S. Mirjalili, Support vector machine: applications and improvements using evolutionary algorithms, in *Evolutionary Machine Learning Techniques*, ed. by S. Mirjalili, H. Faris, I. Aljarah, (Singapore, Springer, 2020), pp. 35–50
29. E. Alpaydin, *Introduction to Machine Learning*, 4th edn. (MIT Press, 2020)

Chapter 11
PDDL Planning and Ontologies, a Tool for Automatic Composition of Intentional-Contextual Web Services

Abdelmajid Daosabah, Hatim Guermah, and Mahmoud Nassar

11.1 Introduction

The exponential use of Web services on the Internet makes them an invocable software component in ubiquitous systems. Thus, the development of explicit methods to describe the semantic aspects of Web services may be influenced by the internal and external context of the execution of the Web service in addition to the desired intention to be achieved by a user. The Web Service Description Language (WSDL) can be used to describe the syntactic interface of a Web service although the combined integration of context and intention as additional descriptions can help in the invocation, discovery, and selection of published Web services. On the other hand, another problem is also raised regarding the search for adequate services that satisfy the multiple needs and expectations of users, which in many cases requires the use of service composition techniques, which are often laborious for the identification and the correct combination of the targeted Web services.

Through this paper, we will present our objective of service composition based on intention and context through the implementation of a practical AI planning technique with Planning Domain Description Language (PDDL) plus the semantic transformation technique adapted from Web Ontology Language (OWL) to PDDL. In addition, we present an overview of our intention-context composition architecture, called CISCA, and the different layers that compose it. Our paper is organized as follows: First, we present works that we consider related to our composition approach. Then we give an overview of the modelling part, which describes the design implemented and the ontologies developed. Then we introduce the AI planning technique with PDDL and the proposed method for mapping it with

A. Daosabah (✉) · H. Guermah · M. Nassar
IMS Team, ADMIR Laboratory, ENSIAS, Mohammed V University, Rabat, Morocco
e-mail: hatim.guermah@um5s.net.ma; mahmoud.nassar@um5.ac.ma

M. Ouaissa et al. (eds.), *Computational Intelligence in Recent Communication Networks*, EAI/Springer Innovations in Communication and Computing,
https://doi.org/10.1007/978-3-030-77185-0_11

OWL. The document ends with a presentation of the CSICA architecture and the layers that compose it, in addition to a conclusion and future work.

11.2 Background and Related Work

In this section, we introduce the concept of context and intention and the value of combining them in a service composition process, noting that we are just going to give some brief snapshots of the definitions and related work that are already presented in detail in another article.

11.2.1 Context and Context Awareness

The notion of context and Context Sensitive Systems that emerged around the 1990s represented a new vision of ubiquitous systems that appeared through the work of Schilit and Theimer [1], Hull [2], and Dey [3], which define context sensitivity as the ability of an application to discover and react to changes in the user's environment. According to Schilit et al. [1], Context Sensitive Systems are defined as systems that adapt to the location of the user and all nearby or accessible people, machines, and devices, as well as to changes over time in these elements.

In other words, it is the consideration of context when implementing a modern mobile and ubiquitous system besides the fact that the mobility of the user provided by the evolution of new technologies has led to the emergence of the context-sensitivity system [4].

Many researchers, such as Schilit and Theimer [1], define the context as a list of the different types of information about the user or the environment in which the application is integrated and which are deemed relevant. These authors limit the definition of context to the observation of the user's location, the set of users present, the objects in proximity, and the changes made to these elements.

Another definition was later proposed by Brown [5]; the author considers that the context includes the location of the user, the identities of the people accompanying him or her, the time of day, the season, the temperature, etc. For Ryan [6], who applies context sensitivity in the case of note-taking software for archaeology, context is more generally defined as the location, environment, identity, and time relative to the user.

In the same perspective, Dey [7] proposes another definition which presents the context as the emotional state of the user, the focus of attention, location and orientation, date and time, objects, and people in the user's environment as the elements constituting the definition of context.

Finally, Chen and Kotz [8] define context as the set of states and environmental parameters that determine the behavior of an application and in which an application event occurs that is of interest to the user.

However, the notion of context has been extended with Mostefaoui [9], Brézillon [10], and Kirsch-Pinheiro [11]; hence, their new definition is no longer limited to the user himself who has performed an action or to his location but extends to physical information such as location, time, etc., and to social and even organizational information such as the role of the user.

We consider that the most illustrative definition of context is the one that includes the notion of context taking into account any information used to characterize the situation; thus, an entity can be a person, a place, or an object considered relevant for the interaction between a user and an application, including the user and his profile and the application itself. This definition of contextual information is grouped into the following categories: individuality, activity, place, time, and relationship [12].

Underlying the various definitions and notions mentioned above, we consider that the context is articulated around the following four specifications [04]:

1. **Context**: It is any information that can be used to characterize the situation of an entity. An entity is a person, place, or object that is considered relevant to the interaction between a user and an application.
2. **Context Aware:** It provides relevant information and/or services to the user, where the relevance depends on the user's tasks.
3. **Interaction with a Context Awareness System**: It can be defined in the executions and configurations. The execution refers to the actions/behaviors of the system in a specific situation (e.g., silence the phone when someone is sleeping). The second refers to the adjustment of actions/behaviors that the system may have. These interactions can be active or passive.
4. **Context Information Life Cycle**: Context has a life cycle, which defines where data are generated and where are consumed [04].

11.2.2 Intention

Taking into account the intention would allow the user's needs to be better satisfied and understand their purpose in a Pervasive Information System (PIS). In an Information System, the main purpose is to answer the user's needs. Expressing these needs in the form of an intention would allow SIPs to understand the "why" of an action, to better assimilate what the user is really looking for, and to better respond to these needs by offering the most appropriate service. Considering this context, we suppose that the service description must consider the concept of intention for a better use of its functionalities whose main objective is to satisfy the intended need of the user.

An intention represents what the user expects when executing a service, representing the user's vision of the functionality he wants in a service [13], which represents the new vision of service deployment and use. According to Rolland [14], this vision helps to bridge the divide between a purely technical and a purely commercial vision of services, focusing on users and their needs. Based on the above, we define the intention as follows (Table 11.1):

Based on Prat [19] and Najar [20] models, we define **intention** as:

- One or more goals to be achieved and accomplished.
- Each goal is represented mainly by two essential clues, a verb and a target:
- A **verb** which exposes the action desired by the user.
- A **target** which satisfies the intention.
- Result created by the action enabling the verb to be realized.
- **Parameters** that contain the additional information needed for the verb being actuated.

Therefore, we consider an intention as follows:

$$\text{Intention} = \sum_{i>=1}^{\text{nGoal}} i = \text{nGoal} * (\text{nGoal} + 1)/2$$

Goal = (verb + target) + [optional constraint (functional or NFunctional)]

11.2.3 Service Composition and Related Work

If the objective of the application designer is not achieved by invoking a single elementary service, then the designer must combine the functionality of a set of services. This process is called service composition [24]. Service composition is not a new subject, so several approaches and frameworks have worked on this subject

Table 11.1 Most appropriate definitions for the intention

Definition	Intention components
An intention expresses a goal, an objective that one wants to attain and that the system must achieve [15].	– Goal – Objective
A goal that the system has to achieve through the cooperation of agents in the future software and in the environment. This author points out a distinction between the notion of intention and that of demand [16].	– Goal – User demand
A prescriptive requirement that must be satisfied by the system under consideration, whereas the requirement is a prescriptive assertion that must be satisfied by the software part of the system only [17]	– Requirement – Satisfaction
The goal that a user wishes to achieve without having to specify how to achieve it, and as the goal to be achieved in order to carry out a process, which is made up of a sequence of sub-intentions and strategies to accomplish it [18]	– Goal – Sub-intention sequence

with different composition methods whether automatic, static, or dynamic and many others that use artificial intelligence for service composition.

In this section, we focus on some of the work that we consider useful and related to our approach.

11.2.3.1 Service Composition Categories

However, before presenting the different composition works, we will first talk in brief about the different categories of composition of services, namely:

- **Automatic composition vs. manual composition**: Manual composition is based on the interaction of the user himself, who plays a pivotal role by manually defining the order in which the component services are invoked and executed; however, automatic composition requires the user to specify his needs in the form of a service request. The system then takes over the entire composition process and carries it out automatically, without any further user intervention.
- **Static composition vs. dynamic composition**: A composition is said static when it takes place at the design stage, when the architecture and design of the software system is planned. The components (or services) that will be used are selected and linked beforehand, and the flow management is carried out a priori [21]. In opposed to static composition, a service composition is said to be dynamic when the services are selected and composed on the fly according to the needs formulated by the user [22].
- **Centralized composition vs. distributed composition**: This category of composition requires a central node that acts as a service composition coordinator and thus imposes a centralized structure. This central node, also called central coordinator or execution engine, takes control of all services involved in a service composition. Its main role is to manage the invocation of the different services involved in the service composition Scheme [23], unlike centralized composition; distributed composition does not have a central coordinator. This type of composition imposes a decentralized structure in which all the services participating in a service composition have to work together to achieve this composition. Thus, the task of carrying out a composite service is distributed among all the services involved.

11.2.3.2 Technical Classification of Service Composition Approaches

In the light of our review of the categories on service composition, we have identified four main approaches, each of it providing a different set of techniques to implement service composition in practice. In order to do this, we illustrate the technical classification of service composition approaches in the table below (Table 11.2):

Table 11.2 Technical classification of service composition approaches

Process composition (workflow): In this approach, the scheme of service composition is defined by specifying the logic of coordination of services through a process. A process, also known as workflow, is represented by an activity-oriented graph and a control flow that gives the order of execution of activities. Each activity represents a functionality provided and performed by a service.	**Service orchestration:** It describes a set of actions in which a given service can or should engage. It offers a centralized vision of the logic of coordination of a composition of services. A central coordinator, called the execution engine, takes control of all the services involved in a service composition and manages the invocation of these different services according to the logic defined by the process. A series of processes executed in a predefined order to achieve a goal
Structural composition (by assembly): The services are provided by components. Each component clearly specifies, in interfaces, the services provided as well as the required services (service dependencies) that are necessary to achieve its business logic. The structural composition thus makes it possible to describe a service composition as an assembly of components. The component assembly consists of configuring and linking the components that provide the services required for the composition.	**Service choreography:** It has no central coordinator. It is interested in how the participating services should work together to achieve a common goal. The logic of the collaboration is divided between the services involved. It is a collaborative effort in which each participating service knows exactly when its operations are to be carried out and with which interaction is to take place.
	OSGi (open services gateway initiative): It is a service platform specification. It offers a service-oriented deployment and execution environment, a component-oriented implementation model, and a number of technical services. The services are implemented and deployed as components called "bundles." A bundle is a deployment unit that corresponds to a JAR file augmented with metadata. This platform allows the deployment and administration of the bundles at runtime. The management of dependencies and interactions between the bundles is done manually by the developer and remains a complex task as it requires a great deal of knowledge of the mechanisms of this approach.
	iPOJO (injected plain old Java object): It is implemented to solve the problem of the complexity of OSGi technology. The main objective is to simplify the development of structural compositions of services, in particular through the implementation of dynamic service links, and automatically manage dependencies and interactions between bundles.

(continued)

Table 11.2 (continued)

	An iPOJO component exposes functionalities across services and expresses dependencies that appear as abstract services to other services, thus delaying the realization of links until execution. The execution of dynamic links is delegated to the container of the iPOJO component. A container is composed of several handlers, each of which takes care of a nonfunctional aspect such as resolving service dependencies, publishing the provided services, etc.
	SCA (service component architecture): It is a service-oriented component model. An SCA component implements one or more services to achieve a certain business logic. An SCA component consists of three parts: The services provided the service references and the associated properties. The composition is achieved by assembling SCA components. These assemblies are called composites. The assembly is carried out at the conceptual level and not at the execution level. The SCA model does not make any proposals for the dynamic adaptation of assemblies to the execution.
Composition by planning based on AI: The composition of services is seen as a problem of planning to achieve a certain objective. This problem has been studied in the field of artificial intelligence (AI). A classical planning problem includes a description of the initial state of the real world, a description of the desired objective, and a description of possible actions that can be carried out. Given a representation of services as actions, the service composition problem comes down to a planning problem so that starting from the user's objective and a set of services, a planner has to find a collection of services that lead to this objective.	**FSA (finite state automata):** It is a class of automata composed of a finite set of nodes representing states, a set of actions, and a set of labelled arcs describing transitions between states. This class of automata is adopted for the specification of service composition processes.
	Situation calculus: A dynamic world is modelled as a world progressing through a series of situations as a result of various actions performed in that world. A situation represents a history of occurrences of actions. The constant "s0" describes the initial situation where no action has yet taken place.

(continued)

Table 11.2 (continued)

	The transition to the successor situation of a situation 's' as a result of an action 'a' is denoted do (a, s). Statements describing a situation are modelled by logical predicate symbols.
	HTN (hierarchical task network): The concept is based on the successive decomposition of tasks. It iteratively decomposes the required task into a set of subtasks until the resulting set contains only atomic (primitive) tasks. The tasks can be executed by directly calling certain atomic operations. At each iteration of the task decomposition, tests are performed to check if certain conditions are not met. The scheduling problem is considered successfully solved if the desired complex task is decomposed into a set of primitive subtasks without violating any of the given conditions.
	Petri nets: It is an oriented, connected, and bipartite graph in which the nodes represent places and transitions, and the tokens occupy places. When there is at least one token in each place connected to a transition, the transition is activated. An activated transition can draw a token from all the entry squares and put it in each exit square. Services are modelled as petri nets by assigning transitions for methods and places for service states. Each service is associated with a petri network, which describes its behavior in this network. A service has two ports: An input port and an output port.
	GraphPlan:
	The service composition problem is represented by a quadruplet (P, W, r_{in}, r_{out}) where "P" represents the set of service input/output parameters, "W" represents the set of services, r_{in} ($r_{in} \subset P$) represents the initial state, and rout ($r_{out} \subset P$) represents the final state (goal). Based on the initial state, it builds a graph of services in several steps called "time steps" in order to achieve the desired goal.

(continued)

Table 11.2 (continued)

	A time step contains two levels: The first level expresses the services whose preconditions are verified and the action of maintaining parameters. The second level expresses the effects of the services satisfied and the parameters maintained. Once the desired goal is reached, GraphPlan proceeds by a backward search to find a valid service composition plan.
	SMA (multi-agent systems): Agents are autonomous entities that seek to achieve a common goal, either through collaboration or with the help of a central entity. Among their advantages is that they can use a common semantic language to support complex interactions. Agents can discover the services they need for invocation either by searching a registry or by asking other agents. Some rules can also be defined by the user to guide the agents' behavior. Agents are not limited to executing a workflow only, but they can follow the user's actions and react to different changes.
Composition by semantic approaches: This category of approaches uses the principles of the semantic web to provide service compositions that take into account the semantic properties of services. While other approaches only identify the structure of the messages exchanged, semantic approaches also interpret the content of these messages. The semantic web proposes to add more information to the description of web services in order to automate several tasks such as service discovery, negotiation, composition, and invocation of services.	**Semantic annotation:** Marks the description of web services with metadata that matches the ontologies. Semantic annotation languages such as DAML+ OIL and OWL are used for publishing and sharing data using ontologies. Semantic web service languages such as OWL-S (formerly DAML-S) provide semantically enriched description for web services.
	Rule-based approaches: They define specific rules that guide the process of composing services. They describe a composability model that verifies which web services can interact with each other according to composability rules that verify the syntactical and semantic properties of the services. Four phases for the automatic composition of services based on the composability model: Specification phase, matching phase, selection phase, and generation phase

(continued)

Table 11.2 (continued)

	Knowledge-based composition: It introduces domain-specific knowledge into the service composition process. This approach is based on the knowledge model of decision support systems, which are based on an intensive use of domain knowledge.
	Input/output dependency: An approach that explores input/output dependencies. Each service is described by certain keywords: The list of its inputs and the list of its outputs. Each output can be linked to either zero, one, or a list of inputs so that the input requirements for generating a specific output may be described. The service composition process begins by creating lists of services that have similar entries and produce similar results. To create a list of services that take some of the initial inputs and possibly output of the services in the previous list, in this way, a service invocation graph will be created.

11.2.3.3 Service Composition Approaches

In this section, we will be describing the various works focusing on service composition, which are closely related to the approach proposed in this paper and which will be reviewed and summarized in comparative Table 11.3 as follows. The main objective of these works is the automation of certain composition-related tasks using semantic information.

11.2.4 Summary

During this section, we have presented some of the several approaches that are close to our proposed vision, showing their strengths and limitations; hence, this works will constitute the pillars of our proposed approach. The next section is devoted to the presentation of the conceptual part of our service composition approach driven by intention and context.

Table 11.3 Comparative table of some composition approaches

Approach/framework/system	Automation	Used technique	Learning	Context	Intention	Advantages	Disadvantages
METEOR-S (managing end-to-end OpeRation): Adapt workflow management techniques for transactional workflows and the integration of semantics in the steps of a service composition [25].	Manual	Workflow	–	–	–	→Consideration of most stages of the life cycle of the composing process	→It does not provide a new model of service representation.
						→The use of semantics in the stages of the composition process	→No main ontology is defined.
							→Supports only OWL
SHOP-2: The main objective is to offer a domain-independent hierarchical planning platform based on the service description in DAML-S for automatic service composition [26].	Auto	HTN	–	–	A given goal	→ Planning in artificial intelligence to create plans by task decomposition.	→ The requirement to have a basic knowledge of the domain
						→ Heuristics for increased performance	→ The difficulty of decomposing activities into sub-activities
							→ Does not support the discovery and selection of services
WSPlan: An approach for the semantic composition of services that focuses on the application of formal reasoning techniques for web services in order to increase the flexibility and autonomy of clients while limiting the complexity and effort of creating such applications [27]	Auto	AI planning	–	–	–	→ Planning and replanning in the event of errors or failures	→Using semantic service markup as a description language, which is not a standardized language
						→ The discovery is made based on reasoning and logical description.	→ Service selection is not supported.
							→ Rescheduling all services in the case of an error

(continued)

Table 11.3 (continued)

Approach/framework/system	Automation	Used technique	Learning	Context	Intention	Advantages	Disadvantages
OWLS-XPlan: A semantic approach for the generation of the service composition automatically using a planner called "OWLS-XPlan" and a semantic information. It takes into account most of the steps of the life cycle of service composition [28].	Auto	HTN	–	–	A goal state	→ Consideration of most life cycle phases in the composition of services	→ Ontology information is not used.
						→Replanning in the case of the failure of one or more services	→Domain-specific knowledge is required.
						→ Use of combined planning techniques	
PORSCE II: The composition process is designed to provide composite services built from a set of available atomic and/or composite services, according to the user's specifications [29].	Auto	AI planning	–	–	User specifications	→ The use of semantic web services to automate discovery and composition	→ Domain-specific knowledge is required.
						→ The use of standards throughout the composition process	

11.3 Intention and Context Modelling

From the approaches presented in the previous section, relating to our vision of intention and context-driven composition, we have identified the focal points to which our approach must respond as well as the methodologies that can be useful in implementing a solution that responds to the expected requirements. As a result, we underline the importance of the solutions presented above, based on service composition and those based on ontologies and rules, as well as on the use of Web services. To guarantee extended use of semantic information and to carry out semantic planning that is sensitive to intentional and contextual constraints, first, we design an abstract and generic meta-model, which expresses the context and intention specifications based on our proposal. Then we will transform our proposed meta-model to an OWL meta-model, using the transformation mechanisms of the ODM model to generate an RDF-/OWL-based ontology model. Finally, we will present the context and intention of ontologies that will be produced from the generated ontology model.

11.3.1 Proposed Meta-Model

To conceive our intention-contextual vision, we propose a generic meta-model whose aim is to guarantee the independence of any specific platform or domain, to be reusable for other extensions if necessary. Figure 11.1 shows the proposed meta-model.

As shown in the figure, this meta-model is based on the following specifications:

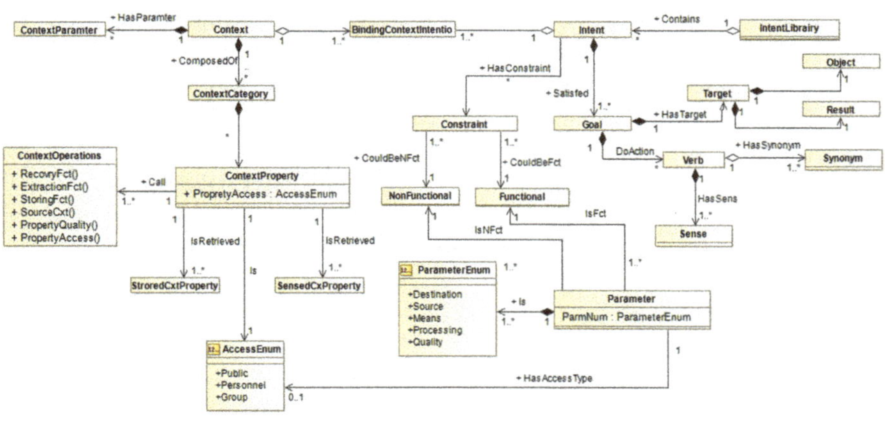

Fig. 11.1 Proposed intention-contextual meta-model

- **Context** is decomposed into a set of sub-contexts that represent the component categories of a given context "ContextCategory."
- **ContextCategory** contains properties that describe it as **ContextProperty** (e.g., quality), so its properties themselves are characterized by:
- **AccessEnum**: types of access to the Context Properties according to the access rights provided (private, personal, group).
- Properties called **StoredCtxProperty** and **SensedCxtProperty**, the first is retrieved from properties already stored, while the second one is retrieved by sensors or different potential sources.
- **Context** has one or more **ContextParameter** that characterize it.
- **Context** can be linked to one or more **Intention** via the **BindingContextIntention**.
- **Intention** seeks to satisfy one or more of the **Goal** that compose the **Intention**.
- **Goal** has one or more objectives to be reached via a "Target" and a "**Verb**" that is presented by an action.
- **Verb** can have at least one **Sense** and **synonym** with the same meaning of the verb.
- **Target** itself responds to an objective and obtains a result.
- "**IntentionLibrary**" contains intentions that are defined by actions characterized by the verb and its corresponding meaning.
- "**Goal**" uses "**Parameter**" (e.g., Source, Quality, etc.) that can be "**functional**" or "**nonfunctional**" (e.g., execution time, the security of the action, etc.).

11.3.2 Ontology-Based Intention and Context Modeller

To respond to our needs for abstract and field-independent modelling that also implements the semantic aspect, we can distinguish two types of techniques, namely, MDA-based approaches and model-driven engineering, which allow the business logic to be independent of the technological platform to avoid frequent changes in technology. However, these approaches do not allow any form of automatic processing of the semantic component. On the other hand, ontologies open up a prospective aspect for context modelling as they allow the classes or categories definition, terms, and relationships between different contextual properties as well as constraints on the use of these classes and properties. On the other hand, an ontology is seen as a semantic network that gathers a set of concepts that describe a specific domain. Thus, the modelling of a concept can change according to the field of study, which poses a great problem in proposing an abstract approach independent of the domain.

To solve the problem of matching ontologies and MDA-based contextual models, we used the OMG standard known as ODM, which allows the modelling of an ontology using an instance of the Meta-Object Facility (MOF) meta-model. Hence, the goal is to allow adding semantics to the already proposed intention-contextual model. The table below shows the correspondence between UML and RDF Schema meta-model. Then the following figure shows the transformation of our view from the intention-contextual meta-model to the OWL language meta-model (Table 11.4).

The Fig. 11.2 describes the following transformation rules:

- **IntToOC** = Intention **To** OWLClass.
- **ICToOP** = Intention Composition **To** OWLObjectProperty.
- **IGToOSP** = Intention Generalization **To** OWLSubjectProperty.
- **ParEnumTo3O** = ParamaterEnum **To** OWL:oneof.

11.3.3 Intention Ontology

According to the above definitions, and taking into account the Prat model [19], we consider that a service satisfies a user's need through the main intention, which seeks to achieve one or more goals. The goal is represented by two main entities, namely, a verb that describes the user's action and a target that satisfies the intention. To enrich the semantics of the intention, we add optional parameters, taken from the model presented by Najar [20], which we categorize as functional constraints influencing the intention's functioning and nonfunctional constraints related to execution functionalities (Fig. 11.3).

11.3.4 Context Ontology

Figure 11.4 illustrates the context ontology after transformation of the intentional part of the previously proposed meta-model into OWL.

Table 11.4 Correspondence between UML meta-model and RDF Schema

UML	Class	Generalization	Association	Attribute	InstanceOf	Attribute type: string	Attribute value
RDFS	Class	subClassOf	Property	Property	Type	Literal	Value

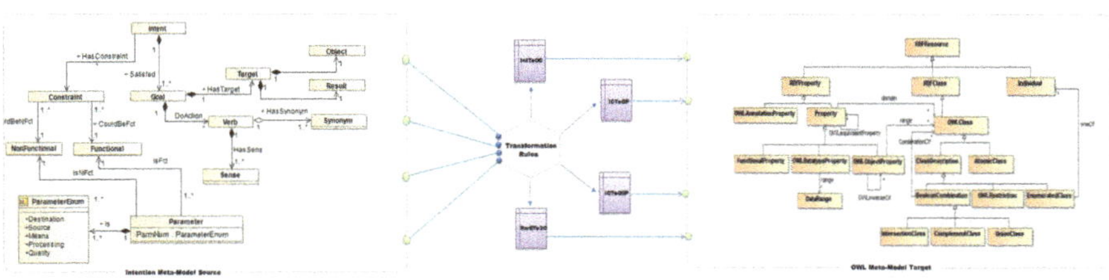

Fig. 11.2 Mapping between the intent-context meta-model to OWL meta-model

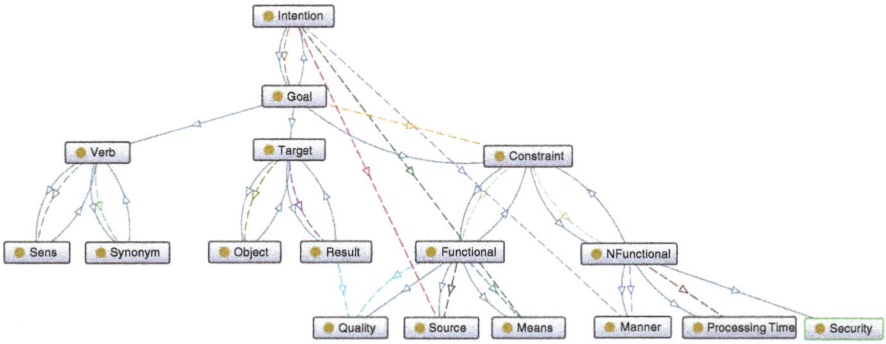

Fig. 11.3 Intention ontology

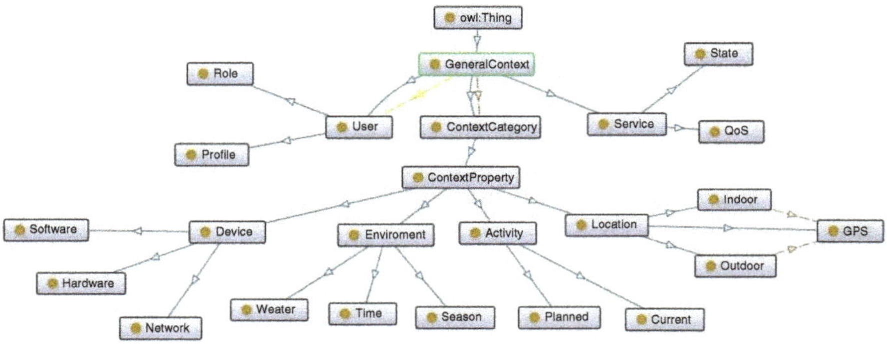

Fig. 11.4 Context ontology

Based on the proposed context ontology, three main classes can be distinguished:

- **User**: Describes a user's profile and role.
- **Service**: Describes the characteristics of a service requested by a user, although its state (Start, In Process, Finish) and quality.
- **ContextCategories**: Describes the belonging of the context to a given category. This context category is characterized by properties that distinguish it, such as the Device, which is a collection of parameters related to memory size, CPU power, bandwidth, battery life cycle, etc.

11.3.5 OWL-S Extension for the Semantic Integration of Context and Intention

OWL-S represents a semantic Web services implementation language dedicated to the description, discovery, selection, invocation service [30], and essentially its composition with the choreography type of Web services. In this section, we focus on the part related to the description of the service composition to extend the OWL-S to consider the user's intention and context. Before discussing the part relating to the OWL-S extension, we remind that a description of OWL-S is composed of three main ontologies:

- **ServiceProfile**: It allows the description and the discovery of services, specifying a description of functional and nonfunctional properties. The Profile part is used for both providers to publish their services and customers to specify their needs (intention). It in turn provides useful information for service discovery and composition.
- **ServiceModel**: This class allows the modelling of services as processes. The subclass defined by OWL-S "Process" is used for service composition; in this sense, the OWL-S models the service as a process. Each process is defined by its inputs/outputs. There are three types of processes in the section on process modelling with OWL-S:
- **AtomicProcess** represents the thinnest level for a process and cannot be decomposed. This means that its execution corresponds to a single step in the service's execution.
- **CompositeProcess** are decomposable into other processes; their decomposition can be specified using a set of control structures such as Sequence, Split, If-Then-Else, etc.
- **SimpleProcess** is not directly invoked. It provides a simplified way to use an atomic service or the means to see a composite process under a single entity.
- **ServiceGrounding:** It indicates how to access the service concretely and provides details about protocols, message formats, and physical addresses.

According to our vision, the extension of OWL-S can be achieved through a modular combination of intentional and contextual ontology, creating a link between the two ontologies and OWL-S. To achieve this goal, we consider an adaptive link to the service, whether composite or simple. This linking mechanism based on the intentional and contextual properties enables the implementation of the necessary conditions and adaptations. Figure 11.5 shows the proposed extension called IACA OWL-S (Intention And Context Adaptation OWL-S).

The extension of OWL-S was inspired by the works of Najar [20] and reflects the intention composition introduced in the works of Kaabi [31] and Rolland [14]. In the abovementioned authors' work, they do not take into consideration the influence of context on obtaining a service that mixes intention and context and focus their actions on intention as an atomic and simple intentional service that can be operationalized through the software service although Najar [20] has worked on this limitation of the reuse of inherited intentional systems and tried to associate these systems with aggregate intentions, without them being able to be assimilated to simple atomic intentions.

To express our vision through the proposed extension of OWL-S, we have implemented certain constraints, which are illustrated in Fig. 11.5, namely:

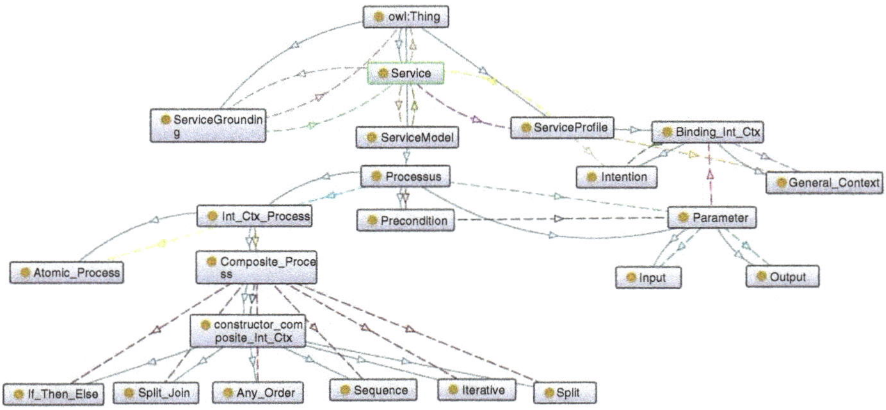

Fig. 11.5 IACA OWL-S

- The **IACA OWL-S** is presented by the classes ServiceProfile, ServiceModel, and ServiceGrounding that are the respective ranges of these properties.
- Each Service instance will present a description of ServiceProfile, will be described by a description of ServiceModel, and will support a description of ServiceGrounding.

"**Binding_Int_Ctx**" is a subclass of "ServiceProfile" that links the modular intentional ontology "Intent" with the modular ontology "**General_Context**. This class is used to retrieve functional and nonfunctional intentional-contextual properties so that they can be exploited for service discovery and composition according to the user's intent.

- "**ServiceModel**" can be seen as a "**Process**," which will be used for service composition.
- "Process" can be conditioned through the "**Precondition**" class, which is parameterized via the "**Parameter**" class linked to the "**Binding_Int_Ctx**" class. This second last one is used to exploit the functional or nonfunctional properties arrived from the "**Binding_Int_Ctx**" class to be used as conditional input parameters.
- "Process" can be seen as an intent-contextual service "**Int_Ctx_Process**"; it can take the form of a simple intent-contextual process performed by an atomic process "**Atomic_Process**" as it can be extended into a composite intent-contextual process "**Composite_Process**," which refines the main intention in the form of sub-intentions which are linked to each other and which satisfy the main goal.
- The processes, as illustrated in Fig. 11.5, link the intentions to each other and to the context through specific "**Constructor_Composite__Int_Ctx**" constructors. These constructors are called constructors of intentional-contextual composition and are inspired by the choice and composition links proposed by Kaabi [Kaabi,

2007]. These links respond to the compositional links that may exist between intention and context.

11.4 PDDL and OWL Interaction

In this section, we will give an overview of AI planning using PDDL, plus the possibility of transforming it into a problem from an OWL, using semantic annotation mapping techniques.

11.4.1 Planning and PDDL

Problem and domain planning is often modelled according to the STRIPS notation (Stanford Research Institute Planning System) [32]. This planning takes the form of a tuple <I, A, G > composed from:

- I, an initial state.
- A is a set of possible and available actions.
- G is a set of objectives to be achieved.
  The states in this planning system are represented as a set of atomic facts, so set A contains all the actions that can be used to modify these states.
- Each action Ai has three lists of facts containing the prerequisites of Ai, Prec(Ai).
- The facts that are added, add(Ai).
- The facts that are removed from the state of the world after the application of the action, Del(Ai).

The following formulae hold for the states in the STRIPS notation [33]:

- An action Ai is applicable to a state S if prec (Ai) $\subseteq$ S.
- If Ai is applied to S, the successor state S′ is calculated as $S' = S - \text{del}(Ai) \cup \text{add}(Ai)$.
- The solution to a planning problem (Plan P) is a sequence of actions P = A1, A2, ..., An, which,
- If applied to I, lead to a state S′ such that $S' \supseteq G$.

The PDDL was designed in 1998 to provide a standard means of encoding planning domains and corresponding problems. It was also used as an entry test for planners who participated in planning competitions. The Planning Domain Definition Language has been improved and extended to become a standard for modelling planning domains and problems. PDDL provides structures to represent all the abovementioned STRIPS elements, such as predicates (atomic facts), actions, and problems. The PDDL also provides separate structures that can be used to represent problems that are associated with specific planning domains. The latest extensions to the PDDL standard take into account the temporal properties of domains, as well as the quality measures, features that could prove very useful in the case of the Web services composition.

11.4.2 Mapping OWL to PDDL

Migrating to an AI planning solution requires the translation of the composition problem of Web service into a planning problem, enriched by semantic data (in our case, intentional and contextual data). This means that the automatic composition of Web services can only be achieved through a semantic annotation, such as preconditions and effects. To this end, we introduce a mapping model between the OWL-S and the PDDL, showing the relevant parts of it through the introduction of a lightweight service description format that allows us to describe the intended semantics of a service.

In this sense, an OWL-S parser deals with the analysis of the profiles of the available atomic Web services and transmits them to the transformation component to translate the semantic descriptions of Web services according to the planning operators. Furthermore, it interacts with the user to formulate the planning problem and deduce both domain and planning problem to the PDDL. Finally and to complete the composition process, the translation operation will be reversed from the PDDL plan to OWL-S (Fig. 11.6).

11.5 Proposed Architecture Overview

In this section, we will give a general overview of our service composition architecture, driven by the user's intention and context, called CISCA (Context-Intentional Service Composition Architecture). The main objective is to show the specification of the necessary meta-models, detailing all the essential concepts related to the development of atomic or composite services that are aware to context and intention. The architecture presents the functionalities offered by each layer and the interactions required between the layers for the composition. Architecture development in layers, by separating the functionalities required by each step of the composition process, has several advantages. First, it allows the framework to be used with multiple technologies, as well as composition methodologies and runtime environments. Second, it also contributes to the isolation of changes from the intention-contextual data, which is communicated to the appropriate layers to handle the sequencing of the composition process.

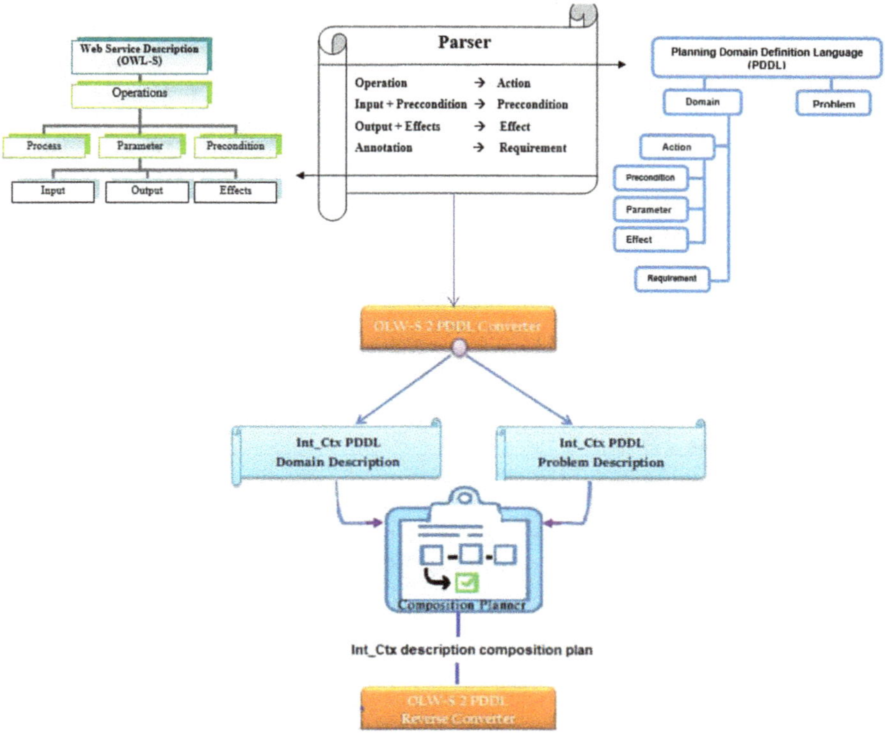

Fig. 11.6 A mapping diagram from OWL-S to PDDL

11.5.1 CISCA Architecture

The CISCA architecture ensures a high-level integration of the composition core, which contains more elaborate relevance indicators, with the possibility of assessing the accuracy of the functions of composite Web services. Besides, CISCA seeks to reduce the complexity of the planning problem generated, through a method of modelling the Web service composition as a planning problem, to speed up the planning process. To guarantee the use of any planning system independently of the domain planners of the CISCA architecture, we propose the inclusion of two external planners. Finally, the composition process is initiated with an OWL-S intention-contextual extension of atomic Web services and concluding it with the OWL-S description of the composite service produced, which makes the deployment process more flexible. Figure 11.7 gives an overview of the architecture and the interactions between its layers.

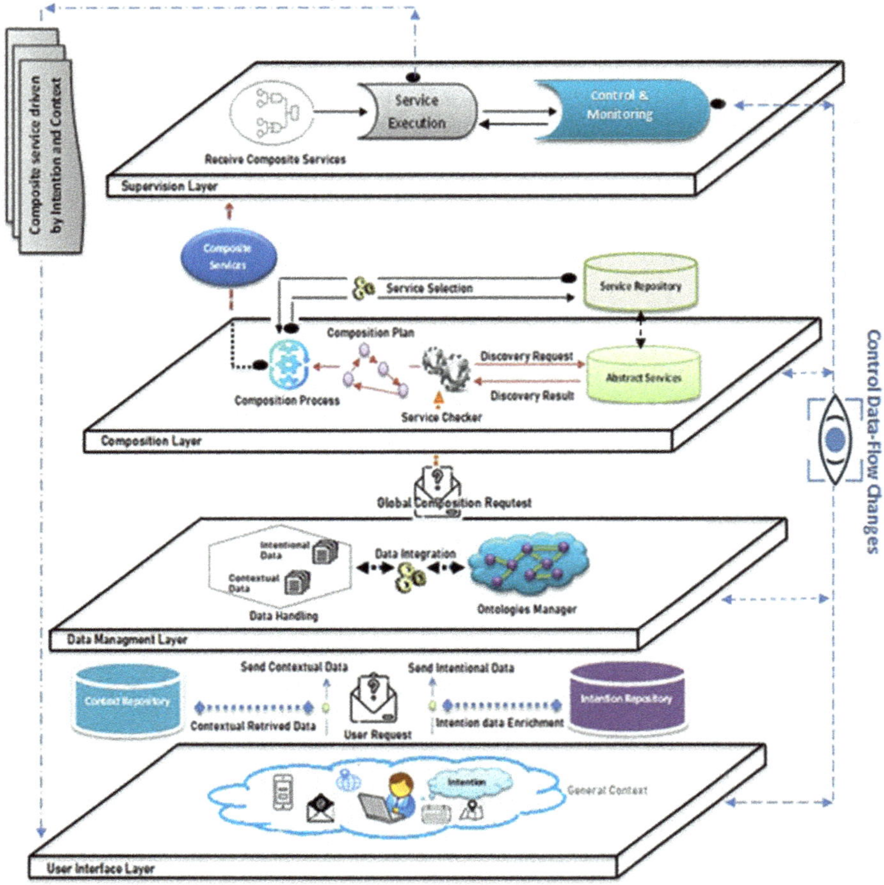

Fig. 11.7 CISCA architecture

11.5.2 CISCA Architecture Features

The CISCA architecture is composed of four levels that are associated with one or more functionalities that they implement. These functionalities are as follows:

1. **User Interface Layer**: Requirement capture, which consists in identifying the different types of users and their context in addition to their needs and the data necessary for the required services, this step is the basis for formulating the contextual and intentional query based on the extracted information.
2. **Data Management Layer**: At the level of this state and from the information extracted by the lower level, the data will be processed through a functional query management mechanism composed of contextual and intentional data, which will dispatch and distinguish the data according to two management mechanisms that operate in parallel, as the first one is used to identify the contextual data retrieved

by comparing them to the existing and stored data; however, the second one seeks to identify the necessary data that expresses the user's intention by classifying the functional and nonfunctional data that determines the goal targeted by the service requester.

3. **Composition Layer**: This layer represents the heart of this architecture, thanks to the "service checker" which represents the orchestral driver between the processed and filtered intention-contextual data and the service repository, selecting from the registered abstract service schemas, those that best match, while determining at the level of the selected abstract services which ones are atomic and which ones will be composed according to a composition scheme, while integrating the functionalities required for the composite service.
4. **Supervision Layer**: This layer represents the final step of this architecture, which consists in delivering the most adequate composite or atomic service that meets the user's requirements according to his context and the parameters integrated with the application of the following sub-functions:

   – Service conformance: Ensures the integrity of a composite service by assessing the compatibility of its description with that of its component services.
   – Service coordination: Controls the execution of the services that take part in a service composition.
   – Service monitoring: Allows observing the execution status of composite services in order to trigger possible service adaptation.
   – Quality of service sensitivity: Verifies that composite services meet QoS (quality of service) requirements based on the QoS provided by their component services.

11.6 Service Composition Module

Composition layer, relevant to the CISCA architecture, contains modules for mapping and preprocessing and planning. In this section, we focus on the service composition module since it represents the key point of this approach. This module consists of the processing of a set of context- and intention-sensitive semantic services, the semantic description according to the OWL-S extension associated with intention-context ontologies, in addition to the composition query. These data are provided as input and a sequence of services, which are pertinent to the definition of the concepts of the domain and the problem to solve. Thus, the architecture of the composition module is as illustrated in Fig. 11.8.

As shown in Fig. 11.8, the composition process module follows the following key points:

- The reception of the global composition request, which contains the semantic intention-contextual data, filtered at the Data Management Layer.
- Mapping the composition request to existing composite services through the cache register.
- The cache register checks the availability of the composite services in case of a match and sends the service composition plan for execution and control.

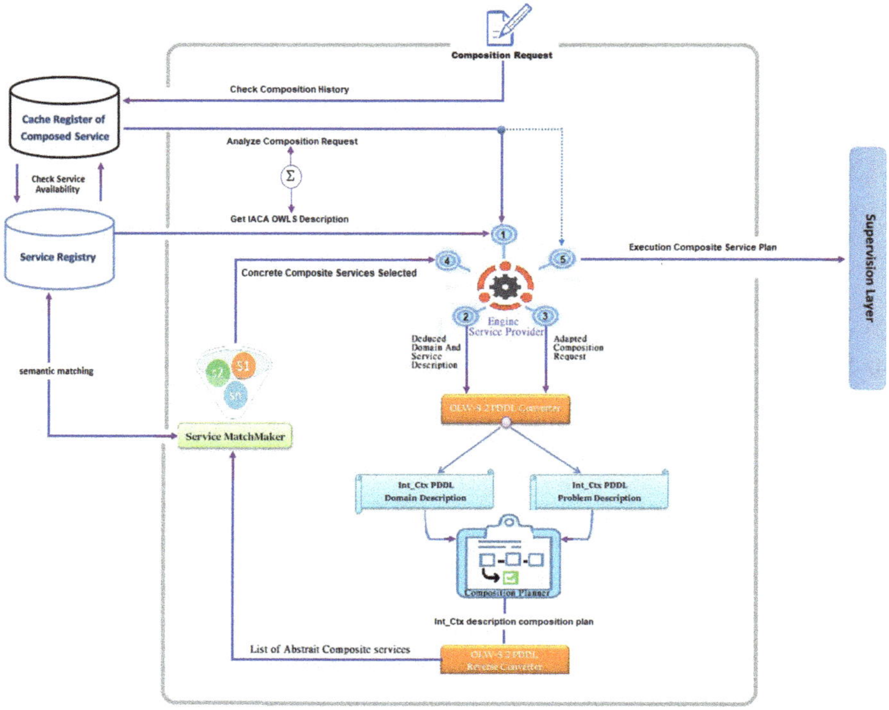

Fig. 11.8 Service composition module

- In case of failed matching with the composition request, a new composition process is started.
- The composition request + IACA are provided as input data.
- Based on the input data, a manager of this process called "Service Provider Engine" handles the adaptation of the composition request and the deduction of the composition domain for the processing of the included object in the request according to a given context.
- The "Service Provider Engine" will apply the mapping technique shown in the figure through an "OWL 2 PDDL Converter," in order to prepare the planning step in the form of PDDL actions and problems.
- The problem (Int_Ctx PDDL Problem Description) and the planning domain (Int_Ctx PDDL Domain Description) obtained by the conversion are managed by the "Composition Planner." It automatically generates the composition plan for the composite service that satisfies the adaptation conditions, according to the intention-contextual objective.
- Once the composition plan has been obtained, the reverse mapping operation is triggered via the OWL 2 PDDL Reverse Converter.
- A "Service MatchMaker" will select the concrete services by applying semantic matching, before applying the composition plan and sending it for execution.

11.7 A Walk-through Example

Waiting for the concrete implementation of our intention- and context-driven composition approach, we are going to set up an example in the form of a table to explain how composite services can be obtained from an intention-contextual query.

We suppose a "remote accident management system," based on the user's intention and context, with the possibility of dynamically adapting the services offered to the different changes in the context or intention of the service requester.

Table 11.5 below maps some possible scenarios, through intention-contextual input parameters, formulated by the Accident system user and the potential outputs of the services (atomic/composite) that can be adapted, after the processing of the user's request containing the context and intention data. The table also shows how

Table 11.5 Example of some possible scenarios to compose the intentional-contextual service

User expression (global request)	User intention (input)	User context (input)	Appropriate service (output)
A deadly traffic accident and a seriously injured person near the agency x in the municipality x	Providing prompt assistance to the injured and transporting the dead	1. Geographic location: ➔ Classified road = No ➔ Unclassified road = No ➔ City area = Yes 2. Damage type: ➔ Material = Yes ➔ Human : ➔Wounded = Yes ➔Deadly = Yes	Civil protection service Hospital service Hearse service Police service Breakdown service
	Action	Object	
	-Prompt assistance -Transport	- Hospital -Mortuary	
A head-on collision between two vehicles on the road X leading back to town Y resulted in serious injuries and one death.	Providing prompt assistance to the injured and transporting the dead	1. Geographic location: ➔ Classified road = Yes ➔ Unclassified road = No ➔ City area = Yes 2. Damage type: ➔ Material = Yes ➔ Human : ➔Wounded = Yes ➔Deadly = Yes	Civil protection service Municipal ambulance service Hospital service Hearse service Police service Gendarmerie service Breakdown service
	Action	Object	
	-Prompt assistance -Transport	- Hospital -Mortuary	

changing the context or intention in the query can influence the service composition process.

11.8 Conclusion and Future Work

During this paper, we introduced the Web service auto-composition tool that we have adopted for our context- and intention-driven approach. To this end, we showed how Web services through their semantic representations could be better exploited by integrating the combination of intention and context in their descriptions. Besides, we have shown the usefulness of using ontologies and how their semantics can be transformed into a problem of Web service composition expressed by PDDL specifications, which are widely supported by a range of planners. We also introduced our adopted CISCA composition architecture and its components plus the composition module and its various key features.

Through this work, we will try in future work to overcome the challenge of the limitations that this PDDL-based composition tool and ontologies may present, namely, the dissociation of composition tasks from the execution of particular planning technologies, i.e., the dynamic choice of the planner best suited to a certain task. In the same sense, we will work on an adaptive composition algorithm, to integrate context and intention, which will be supported by the planner during the composition process.

References

1. B.N. Schilit, M.M. Theimer, Disseminating active map information to mobile hosts. *IEEE Network: The Magazine of Global Internetworking*, (1994)
2. R. Hull, P. Neaves, J. Bedford-Roberts, Towards situated computing, in *First International Symposium on Wearable Computers (ISWC)*, (1997), pp. 146–153
3. A.K. Dey, Providing architectural support for building context-aware applications. Georgia Institute of Technology. Thèse de doctorat, Georgia Institute of Technology, (2000)
4. A. Daosabah H. Ghermah, Dynamic composition of services: an overview of approaches led by the context and intent of the user. The 4th International Conference On Big Data and the Internet of Things – BDIoT'19, (2019)
5. P.J. Brown, J.D. Bovey, X. Chen, Context-aware applications: from the laboratory to the marketplace. IEEE Pers. Commun. **4**(5), 58–64 (1997)
6. N. Ryan, J. Pascoe, D. Morse, Enhanced reality fieldwork: the context-aware archaeologist assistant, in *Computer Applications & Quantitative Methods in Archaeology*, Exon, ed. (1997)
7. A.K. Dey, G.D. Abowd, The context toolkit: aiding the development of context-enabled applications, in The *Proceedings of the SIGCHI conference on Human Factors in Computing Systems*, (1999), pp. 434–441
8. G. Chen, D. Kotz, A Survey of Context-Aware Mobile Computing Research. TR2000-381. Dept. of Computer Science, (Hanover: Dartmouth College, 2000)
9. G.K. Mostefaoui, J. Pasquier-Rocha, P. Brézillon, Context-aware computing: a guide for the pervasive computing community, in *Proceedings of the IEEE/ACS International Conference on Pervasive Services (ICPS)*, (2004), pp. 39–48

10. P. Brézillon, Task-realization models in contextual graphs, in *Modeling et Using Context, in 5th International et Interdisciplinary Conference CONTEXT, 3554 of Lecture Notes in Computer Science*, (Springer Verlag, 2005), pp. 55–68
11. M. Kirsch-Pinheiro, Adaptation Contextuelle et Personnalisée de l'Information de Conscience de Groupe au sein des Systèmes d'Information Coopératifs. Thèse de doctorat, Université Joseph-Fourier – Grenoble I (2006)
12. A. Zimmermann, A. Lorenz, R. Oppermann, An operational definition of context, in *Modeling and Using Context*, (Springer Berlin Heidelberg, 2007), pp. 558–571
13. D. Fensel, F.M. Facca, E. Simperl, I. Toma, *Semantic Web Services* (Springer, Berlin, 2011), 350 p
14. C. Rolland, M. Kirsch-Pinheiro, C. Souveyet, An intentional approach to service engineering. IEEE Trans. Serv. Comput. **3**(4), 292–305 (2010)
15. V. Plihon, J. Ralyte, A. Benjamen, N.A. Maiden., A. Sutcliffe, E. Dubois, P. Heymans, A reuse-Oriented Approach for the Construction of Scenario Bases Methods, in *Proceedings of International Conference on Software Process* (1998)
16. A. Van Lamsweerde. Requirements engineering in the year: a research perspective, in *Proceedings of the 22nd International Conference on Software Engineering*, (New York, NY, USA:ACM, 2000), pp. 5–19
17. A. Van Lamsweerde, Goal-oriented requirements engineering: a guided tour, in *Proceedings of the Fifth IEEE International Symposium on Requirements Engineering*, (2001), pp. 249–262
18. R.S. Kaabi, C. Souveyet, Capturing Intentional services with Business Process Maps, in *1rst IEEE, International Conference on Research Challenges in Information Science (RCIS)*, (2007), pp. 309–318
19. N. Prat, Formalisation et classification des objectifs pour l'ingénierie des exigences. Dans Proc. Du 3ème atelier international sur l'ingénierie des exigences: fondements de la qualité logicielle (1997)
20. S. Najar, M. K. Pinheiro, Y. Vanrompay, L. Angelo, C.S. Steffenel, Intention Prediction Mechanism In An Intentional Pervasive Information System. Intelligent Technologies and Techniques for Pervasive Computing, IGI Global, (2013), pp. 251–275
21. S. Dustdar, W. Schreiner, A survey on web services composition. Int. J. Web Grid Serv. **1** (2005)
22. T. Osman, D. Thakker, and D. Al-Dabass. Bridging the Gap between Workflow and Semantic-based Web services Composition, in *Proceedings of the Web Service Composition Workshop WSCOMPS05*, (2005)
23. A. Yachir, Thèse de doctorat « composition dynamique de services sensibles au contexte dans les systèmes intelligents ambiants », Université Paris-Est (2014)
24. B. Benatallah, R. Dijkman, M. Dumas, Z. Maamar, Service composition: concepts, techniques, tools and trends, in *Service-Oriented Software Engineering: Challenges and Practices*, ed. by Z. Stojanovic, A. Dahanayake, (Idea Group Inc (IGI), 2005), pp. 48–66
25. R. Aggarwal, K. Verma, J. Miller, W. Milnor, Constraint driven web service composition in METEOR-S. In Services Computing, 2004. (SCC 2004). Proceedings. 2004 IEEE International Conference on, IEEE, (2004) pp. 23–30
26. K. Erol, J. Hendler, D.S. Nau. HTN Planning: complexity and expressivity, in *Proceedings of the 12th National Conference on Artificial Intelligence (AAAI 1994)*, (AAAI Press, Seattle, WA, USA. July-Aug. 1994) pp. 1123–1128
27. J. Peer, Description and Automated Processing of Web Services. PhD thesis, the University of St. Gallen, Graduate School of Business Administration, Economics, Law and Social Sciences (HSG), (2006)
28. M. Klusch, A. Gerber, M. Schmidt. Semantic web service composition planning with owls-xplan, in *Proceedings of the 1st Int. AAAI Fall Symposium on Agents and the Semantic Web*, (2005), pp. 55–62
29. O. Hatzi, M. Nikolaidou, D. Vrakas, N. Bassiliades, D. Anagnostopoulos, I. Vlahavas, Semantically aware web service composition through AI planning. Int. J. Artif. Intell. Tools **24**(01), 1450015 (2015)

30. D. Martin, M. Burstein, D. Mcdermott, S. Mcilraith, M. Paolucci, K. Sycara, D.L. Mcguiness, E. Sirin, N. Srinivasan, Bringing semantics to web services with OWL-S. World Wide Web **10**(3), 243–277 (2007)
31. R.S. Kaabi, C. Souveyet, Capturing Intentional services with Business Process Maps, in *1rst IEEE, International Conference on Research Challenges in Information Science (RCIS)*, (2007), pp. 309–318
32. Fikes, Nilsson, Strips: a new approach to the application of theorem proving to problem solving Presented at the 2nd IJCAI, Imperial College, London, England, September 1–3, (1971)
33. O. HAtzi, D. Vrakas, N. Bassiliades, D. Anagnostopoulos, The porsce II framework: using AI planning for automated semantic web service composition, (2010)

Chapter 12
QSAR Anti-HIV Feature Selection and Prediction for Drug Discovery Using Genetic Algorithm and Machine Learning Algorithms

Houda Labjar, Najoua Labjar, and Mohamed Kissi

12.1 Introduction

The process of drug discovery and production is an inherently long and difficult undertaking, with the development of a compound from initial discovery to market launch for at least 10 years. It is recognized as a costly undertaking although the precise costs and the methodology to calculate them have been much debated. Several antiretroviral (ARV) drugs have been approved by the food and drug administration for use in the therapeutic treatment of human immunodeficiency virus (HIV) infection. Most of the time, anti-HIV drugs are used with other drugs given together, usually involving two reverse transcriptase inhibitors and one inhibitor of one of the other classes of drugs. There is a significant need for the ARV therapy described above. At the end of 2019, an estimated 36 million people (including 34 million in low- and middle-income countries) were infected with HIV, and 1.7 million new infections occur each year [1]. The sub-Saharan Africa region most affected by the HIV/acquired immunodeficiency syndrome (AIDS) epidemic, it accounts for 67% of people infected with HIV globally (around 24 million people) and with almost 72% of all deaths due to AIDS per year.

H. Labjar
Faculty of Sciences and Technology, University Hassan II Casablanca, Mohammedia, Morocco

N. Labjar
CERNE2D, ENSAM, University Mohamed V, Rabat, Morocco
e-mail: najoua.labjar@um5.ac.ma

M. Kissi (✉)
Advanced smart systems, LIM Laboratory, Computer Science Department, Faculty of Sciences and Technology, University Hassan II Casablanca, Mohammedia, Morocco
e-mail: mohamed.kissi@fstm.ac.ma

M. Ouaissa et al. (eds.), *Computational Intelligence in Recent Communication Networks*, EAI/Springer Innovations in Communication and Computing,
https://doi.org/10.1007/978-3-030-77185-0_12

The main focus of the present work is to develop a quantitative structure-activity relationships (QSAR) model able to correlate the structural features of the 1-[(2-hydroxyethoxy) methyl]-6-(phenylthio)thymine (HEPT) derivatives with their anti-HIV activities. QSAR models have been proposed to estimate biological property, such as activity [2], with the aim of providing adequate and relevant information for drug discovery [3].

In general, QSAR models require the representation of the chemical, physical, and electronic structure of compounds by their molecular parameters [4], such as the constitutional, geometric, and quantum descriptors. The development of a QSAR model consists of calculating all molecular descriptors, but only a subset of these descriptors provides the information needed to obtain the powerful predictive model [5]. Subsequently, obtaining the powerful predictive models is based on the good selection of the molecular descriptors used during the generation of the QSAR model [6].

Several machine learning algorithms have been used for processing the molecular descriptors selection, which are known in informatics as feature selection methods. Generally, the used approaches are divided into three types: filter, wrapper, and embedded methods. A disadvantage of the filter approaches is that they examine each feature independently and ignore the individual performance of the feature in relation to the all-group features [7]. This problem can affect the machine learning results used below. For wrapper and embedded methods, the machine learning algorithms require an implementation heavy in time complexity, which constitutes an important problem.

The genetic algorithm (GA) is a method that can be used for feature selection [8]. Often, feature selection methods using GA requires a classification machine learning algorithm as support vectors machines (SVM) [9], artificial neural networks (ANN) [10], k-nearest neighbor (KNN) [11], decisions trees (DT), and random forest (RF) [12], to evaluate each individual in the population, and continues with other GA operators as selection, crossover, and mutation.

In this work, a new approach feature selection based on GA for QSAR anti-HIV is proposed, which incorporates with many machine learning algorithms for each iteration in this GA. The individual evaluation is performed in each iteration using the same classification machine learning algorithm. The fitness function is represented through the classification error found by the machine learning algorithm associated with features selected with the GA. More specifically, the specified machine learning algorithm is formed for each iteration before GA starts to search new features set by improving the classification model error. This approach will be tested on the anti-HIV QSAR problem, and then it will be compared with the new approaches in the literature dealing with the QSAR modeling problem.

The structuration of the rest of the paper is as follows: In Sect. 12.2, the techniques of feature selection is briefly motivated by presenting feature selection problem in general and in QSAR problem. Used data and the proposed approach are presented in Sect. 12.3 based on GA and machine learning algorithms, and the experimental results with analysis and discussion are presented also in this section, and this work is concluded in Sect. 12.4.

12.2 Proposed Techniques

The feature selection is a technique for choosing characteristics, variables, and attributes that are more interesting, relevant, and informative for supervised classification problem as well as unsupervised. It consists of choosing from a set of large variables, a subset of interesting features for the studied problem. The application fields of feature selection techniques are varied, for example, modeling areas, classification, machine learning, and data mining, but the aim of this work is to propose an approach for the feature selection in the supervised classification case in QSAR problem. In this case, the purpose of feature selection is to find an optimal subset of features, molecular descriptors, that are relevant and nonredundant. In addition, this subset should satisfy accuracy and rapid learning or yet the applicability of the proposed QSAR model classification. That is to say that the subset must be confirmed by the chemist human expert.

QSAR models establish the relationship between structure and activity of chemical or biological compounds [13]. The process, for achieving QSAR models, is based on supervised machine learning algorithms. These models have the ability to classify correctly the compounds forming the training database and to predict the activity of newly synthesized compounds [14]. The realization of these models begins with a selection step of the most relevant molecular descriptors for the modeling of the activity target [15], which constitutes one of the characteristic selection problems. A training database of compounds is used to carry out this molecular descriptor selection process. This training database includes compound examples with their molecular descriptors and the measured activity of each compound. The machine learning strategy consists in selecting and evaluating the different descriptors in order to identify the smallest and best subset of these molecular descriptors.

12.2.1 Genetic Algorithms

The GAs are one of the evolutionary-based techniques that are used nowadays intensively in the feature selection field. The application of GAs, to solve an optimization problem, requires encoding all potential solutions to this problem in chromosome forms. This is to find a good selection function for discrimination between the chromosomes set using genetic operators. The GAs are stochastic optimization algorithms that are based on the mechanisms of natural selection and genetic [16]. The algorithm started with an initial chromosome population arbitrarily chosen, and the performance or fitness of each chromosome is evaluated. A genetic algorithm is an iterative algorithm for optimum search; it manipulates a population of constant size. Constant population size causes a phenomenon of competition between chromosomes. Each chromosome represents the coding of a potential solution to the asked problem; it consists of elements called genes.

For each iteration, called generation, a new population is created with the same number of chromosomes. This generation involves better chromosomes adapted to their environment using selection function. As the generations, the chromosomes will tend toward the optimum selective function. The new population creation from the previous one is based on the genetic operator's application to know selection, crossover, and mutation. Recently, many works in pattern recognition systems used evolutionary algorithms for feature selection. In this sense, genetic algorithms method has proven effective as feature selection technique QSAR studies [17].

12.2.2 Machine Learning Methods in QSAR Problem

Machine learning techniques can be classified as supervised or unsupervised learning. For supervised learning, labels are assigned to the training data, and once trained, the model can predict labels for data entries. Supervised machine learning models include regression analysis (RA), artificial neural networks (ANN), k-nearest neighbor (kNN), decision trees (DT), random forests (RF), Bayesian probabilistic learning (NB), and support vector machines (SVM). Unsupervised machine learning techniques directly learn the underlying patterns of molecular characteristics from unlabeled data. Clustering algorithms represent a family of unsupervised algorithms, where the data set is first divided by predefined distance measures in a high-dimensional space, and the labels are then assigned based on the number of categories observed. Recent machine learning algorithms provide a powerful suite of techniques for finding and identifying nonlinear QSAR relationships with greater precision and accuracy.

ANNs have been applied to QSAR modeling properties. Among the latest work is that of Baskin et al. Their work provides an approach based on ANN in chemoinformatics [18]. A new QSAR model, using the naïve bayes classifier, is presented by Pradeep et al. [19]. The model makes it possible to vary a cutoff parameter which allows a selection in the desirable compromise between the sensitivity and the specificity of the model. The results of the cross-validation show that the model achieves the best compromise between sensitivity and specificity in QSAR problem prediction. The kNN algorithm is also applied in QSAR problem. The best QSAR model was obtained using kNN algorithm with $k = 5$, which gave an overall accuracy of 76.6% for the training data and 77.9% for the test data [20]; however, these classification rates are not satisfactory enough. Kyoungyeul et al. has developed a QSAR model using a random forest algorithm [21]. The recall rate of the highest targets is calculated to assess the performance of the ranking of targets. This rate is on the order of 73.9%, which remains a non-sufficient rate for the case of the QSAR problem. Recently, the SVM algorithms are introduced to solve the QSAR problem and the complex nonlinear relationship between molecular descriptors and activity of compounds that have a similar structure [22].

12.3 Proposed Approach and Validation

12.3.1 Data

In the experiment's setups, the data set of recognition of anti-HIV activity is used. This data set consists of 81 anti-HIV molecules with their inhibitory activity. It is taken from articles published by Tanaka et al. [23] and Garg et al. [24]. The anti-HIV activity of the compounds has been expressed by the ability of the compound to protect cells against the cytopathic effects of the virus. This activity, the concentration needed for 50% effect, was measured in MT-4 cells cultures and expressed by *pIC50*. The chemical data structures used in this work are those used in the work of Bazoui et al. [25].

The objective of QSAR studies is to search the relationship that exists between the structure of molecules and their activities, in particular the study activity anti-HIV. A major step in all QSAR studies is to select and compute molecular descriptors as coded numeric variables representing chemical, physical, geometric, and electronic structures. As all the compounds studied have the same basic skeleton, each molecule is described by means of properties of the substituents, linked to the basic skeleton. Determining the relevant properties for a given substituent is useful for assessing local interactions between the molecule and the receptor site. In this work, the properties of the molecule that looked are its size, its height, its molecular weight, its lipophilicity, etc. All these properties have been measured to know the possibility of being transported, accessed, and interacted with the given receiving site. These descriptors show the hydrophobic, steric, geometric, and electronic aspects.

The molecule descriptors used are:

- *LogP*: Logarithm of the partition coefficient between the water and the octanol of the molecule
- X_1: fourth-order connectivity index
- X_2: sixth-order connectivity index
- *S*: Surface of the molecule
- *V*: Volume of the molecule
- B_1, B_2: The parameters of the substituents of the molecule
- *MR*: Molecular reactivity
- *MW*: Molecular weight
- *Ov*: Ovality estimation
- *L*: Length is the distance along the screen x-axis between the left- and rightmost atoms
- *W*: Width is the distance along the screen y-axis between the top- and bottommost atoms
- *H*: Height is the distance along the screen z-axis between the nearest and farthest atoms
- The ratios V/L, V/W, and W/H were also calculated

The number of all descriptors or variables for each molecule is 16 and calculated using Molecular Modelling Pro software (MMP) [26].

12.3.2 Proposed Approach

To select the most important descriptors, an approach has developed that combines genetic algorithm and many machine learning algorithms (ANN, KNN, NB, RF and SVM). This approach was applied using the training data set formed by 16 attributes (molecular descriptors) and the label *pIC50* (activity). Each training set example described by values for these attributes is associated with a label value of *pIC50*. The aim is to find the relevant descriptors which better explain the activity.

The approach begins with research of the two most important selected descriptors using the GA algorithm. These descriptors constitute the attributes of the training data set inputs to develop the machine learning QSAR model with predicting the *pIC50* label.

The steps of the GA-machine learning approach (GA-MLA) are as follows:

Step 1. GA-MLA starts with a randomly initial population of 200 individuals. Each individual is composed with $T = 2$ and two descriptors D_1 and D_2 from the 16 descriptors used in the initial data set. T is the number of molecular descriptors constituting an individual population.

Step 2. Application of machine learning algorithm for all individuals. 200 QSAR models are obtained.

Step 3. Selection operator: Randomly 100 individuals are chosen (from the 200 in Step 2) with a good accuracy of prediction of all 81 molecules.

Step 4. Crossover operator: The crossover is applied in two parents. Each parent is composed by two descriptors, Parent 1 (D_i, D_j) and Parent 2 (D_k, D_l) selected randomly from Step 3. The crossover operator permutes the first component of Parent 1 (D_i) and the second component of Parent 2 (Dl). Two new children, each child is composed by two descriptors. Children 1 and Children 2 are created (see the crossover step in Fig. 12.1). This step generated 100 new individuals.

Step 5. Mutation operator: The mutation operator is applied to 20 individuals selected randomly from Step 4 (the mutation probability is equal to 0.25). This operator changes a descriptor of the two descriptors for the 20 individuals selected for this operator. For example, in the individual (D_i, D_j), the descriptor D_j is changed by D_h, $1 \leq \text{h} \leq 16$, and h is different to i. The individual becomes (D_i, D_h) (see the mutation step in Fig. 12.2).

Step 6. Then Step 3, Step 4, and Step 5 generate 100 new individuals which will be added with the 100 individuals rested in Step 2. An update population is generated. Steps 2 to Step 5 are repeated. A certain number of iterations are tested. The algorithm stops when there is no change for the next population. The two relevant molecular descriptors are obtained for the end of GA-MLA. The GA-MLA model flowchart is shown in Fig. 12.1.

Data set preparation

N°	X_1	B_2	*LogP*	*S*	...		*pIC50*
1	0.26	1.52	1.87	1.55	...		4.15
2	0.27	1.7	1.11	1.57	...		3.87
3	0.27	1.35	0.89	1.65	...		4.72
4	0.27	1	1.87	1.63	...		5.59
5	0.26	1	2.4	1.71	...		5.57
..	..	..	...	...	...		...
...	...	...	...	...	...		...
...	...	...	...	...	...		...
...	...	...	...	...	...		...
79	0.29	1	3.04	1.59	...		7
Molecular descriptors							Activity

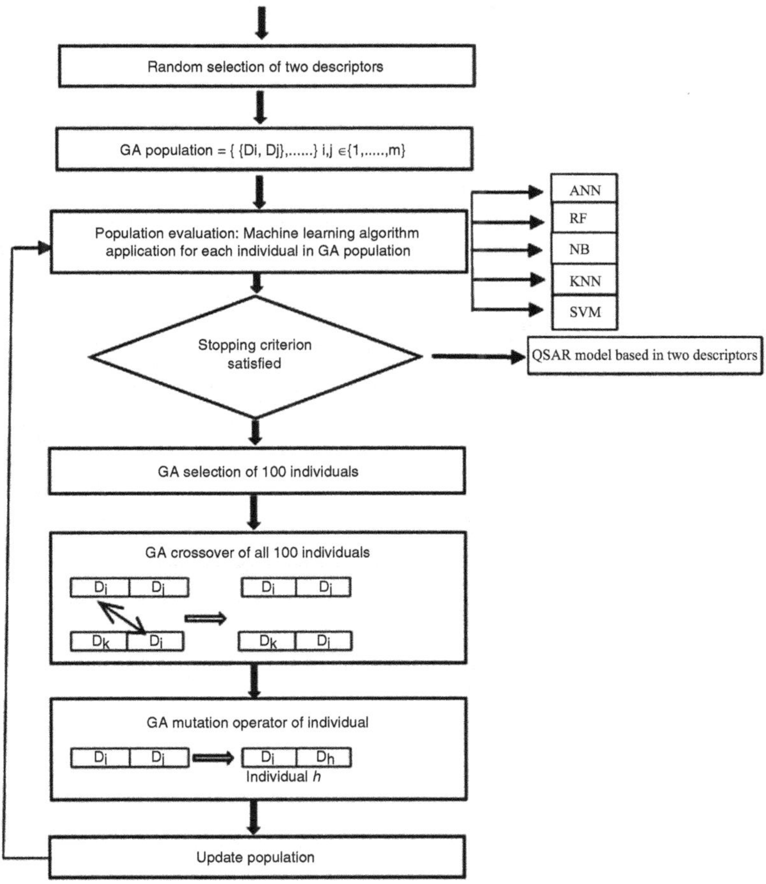

Fig. 12.1 The proposed GA-MLA approach flowchart

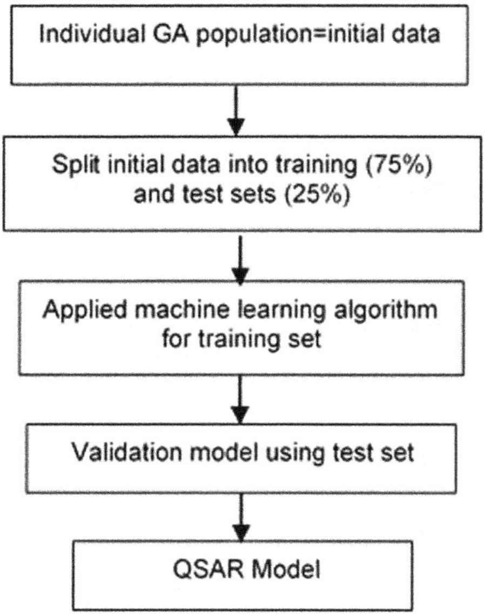

Fig. 12.2 Population evaluation in GA-MLA approach

The population evaluation in Step 2, for each individual GA population, with machine learning algorithm is presented in Fig. 12.2. The initial data set, constituting the GA population, will be divided into two subsets. The first data subset is the training data set. It is on the order of 75% of the total data set. The second data subset, the rest of the total data set is around 25%, is the test data set of the model using the machine learning algorithm. The best QSAR model, based on the two pertinent descriptors molecular, is generated.

The individual's population of the genetic algorithm change their size by adding randomly another molecular descriptor each time. The genetic algorithm shown in Fig. 12.1 is performed again at an individual population of size T, which is equal to three, four up to 16 molecular descriptors (16 is the total number of molecular descriptors). The algorithm process is applied again to a population of the genetic algorithm, of individuals whose size of each individual is T descriptors ($3 \leq T \leq 16$). The case of T equals 2 is the operation of flowchart showed in Fig. 12.1. The best QSAR model of each T is selected ranging from 2 to 16 molecular descriptors, and then the final QSAR model is the best of these 15 QSAR models. This final QSAR model gave the relevant T molecular descriptors that better explain the *pCI50* activity.

12.3.3 Results and Validation

In this section, the application of the proposed approach is described. The results of the evaluation are given, which concerns the number of selected features or descriptors *T*, the specificity, sensibility, and accuracy rates.

Each example, of training data set, containing values for these molecular descriptors, is associated with a class value (anti-HIV). Consider an instance (example) of the data set studied. Its description is associated with the value *pIC50* (anti-HIV) = 3.87 to obtain an example of learning with all molecular descriptors (see Table 12.1). Once all the examples (molecules) are described by all different molecular descriptors, the label or class is coded by the anti-HIV activity.

The proposed genetic algorithm will increase the number of descriptors for each iteration and look for the best features using machine learning algorithm. After a certain number of generations fixed at the beginning, the results found are recapitulated in Table 12.2.

For each selected molecular descriptors, sensitivity, specificity, and accuracy are showed of the obtained classification result. Accuracy is calculated as the ratio between the number of correctly classified molecules and the total number of molecules. Sensitivity and Specificity are defined and calculated as follows:

Table 12.1 Description of an instance example

X_1	B_1	V	X_2	Ov	B_2	L	MW	W	H	V/L	V/W	W/H	$LogP$	MR	S	pIC50
0.26	1	1.55	0.042	0.58	1.52	2.25	250	4.5	3.25	0.69	0.34	1.38	1.87	0.57	0.65	3.87

Table 12.2 Evaluation of classification performance for each selected molecular descriptor

QSAR model	Machine learning algorithm	*T* most descriptors found by GA	Error	Sensibility (%)	Specificity (%)	Accuracy (%)
1	SVM	X_1, S	0.13	93	82	88
2	ANN	L, H	0.235	83	72	78
3	SVM	$S, B_2, X_2, LogP, X_1$	0.125	95	83	89
4	ANN	$S, X_1, X_2, LogP, L$	0.145	89	79	85
5	SVM	$S, X_1, X_2, LogP, L, B_1$	0.34	77	64	67
6	RF	V, H, X_2, L, MR, B_2	0.38	75	61	63
7	NB	$X_1, B_2, X_2, logP$	0.2	84	73	79
8	RF	L, V/W, X_1, $logP$	0.29	77	71	72
9	SVM	$X_1, logP, S$	0.03	99	91	97
10	KNN	X_1, V, W	0.27	82	73	76

$$Sensibility = \frac{TP}{TP + FN}$$

$$Specificity = \frac{TN}{TN + FP}$$

where *TP* (true positive) is correctly anti-HIV, *FP* (false positive) is incorrectly anti-HIV, *TN* (true negative) is correctly anti-HIV, and *FN* (false negative) is incorrectly anti-HIV.

Following the results in Table 12.2, it is noted that the features number T for most descriptors is reduced by improving the accuracy, sensibility, and specificity rates. On the other hand, it is remarkably the efficiency of the propose approach to select the most relevant features after selecting the following three features ($T = 3$): X_1, *LogP*, and *S* in the nine QSAR model using SVM algorithm.

The best machine learning algorithms, found by the genetic algorithm, with the best selected molecular descriptors are represented in Table 12.3.

The nine QSAR model analysis, in Table 12.2, was performed using selected molecular descriptors (X_1, *logP*, *S*) and the experimental values of the anti-HIV activities for the 1-[(2-hydroxyethoxy) methyl]-6-(phenylthio). The values of calculated activities using the proposed approach, based on the genetic algorithm and SVM as machine learning algorithm (GA-SVM), have been presented in Table 12.4.

The predictive capacity, of the model found, is to test their ability to perfectly predict the *pIC50* of the compounds from an external data test set (25% of the compounds that were not used for the model constructed). The *pIC50* of the remained set of 21 compounds are calculated using the proposed QSAR model with the compounds used in the approach with GA and SVM. The observed and calculated *pIC50* values are given in Table 12.5. This model is able to predict the activities of the molecules of the data test set in comparison with the experimental values of these activities. The correlations of calculated and observed activities values *pIC50* for the data test set are illustrated in Fig. 12.3.

For more investigation, artificial neural networks (ANN) [27], support vector machines (SVM) [28], multilinear regression (MLR) with SVM [29], and deep learning with long short-term memory neural networks (LSTM-NN) [30] techniques, with all chemical descriptors, are used to predict the anti-HIV activity.

Table 12.3 Selected descriptors with the best machine learning algorithm accuracy

T descriptors found by GA	SVM	KNN	ANN	RF	NB
X_1, S	**88%**	81%	82%	79%	83%
S, X_1, X_2, *LogP*, L	80%	79%	**85%**	78%	81%
X_1, *logP*, S	**97%**	91%	93%	87%	89%
S, B_2, X_2, *LogP*, X_1	**89%**	83%	88%	80%	85%

Table 12.4 Observed and predicted activities using the QSAR model based on GA and SVM for the training set

Data no.	Observed pIC50	Predicted pIC50	Residues	Data no.	Observed pIC50	Predicted pIC50	Residues
1	4.15	4.15	0.00	41	5.06	5.06	0.00
2	3.87	3.85	0.02	42	5.17	5.17	0.00
4	5.59	5.60	0.00	44	6.48	6.30	0.18
5	5.57	5.57	0.00	45	5.82	5.82	0.00
6	4.92	4.92	0.00	46	5.24	5.24	0.00
7	4.35	4.40	−0.05	48	5.48	5.48	0.00
8	5.48	5.48	0.00	49	7.06	7.06	0.00
10	5.24	5.24	0.00	50	7.72	7.72	0.00
11	5	5	0.00	51	7.58	7.58	0.00
12	4.47	4.52	0.05	53	8.3	8.3	0.00
14	4.66	4.66	0.00	55	8.55	8.55	0.00
15	6.59	6.59	0.00	56	8.09	8.09	0.00
16	5.89	5.89	0.00	57	8.14	8.20	−0.06
17	6.66	6.66	0.00	58	7.99	7.99	0.00
18	5.1	5.2	−0.1	60	7.89	7.89	0.00
20	5	5	0.00	61	8.14	8.10	0.04
22	6.96	6.96	0.00	62	5.68	5.68	0.00
23	5	5	0.00	64	5.66	5.66	0.00
25	8.11	7.97	0.13	66	7.89	7.89	0.00
26	8.3	8.3	0.00	67	6.66	6.66	0.00
27	7.37	7.37	0.00	68	5.79	5.91	−0.12
29	5.47	5.47	0.00	69	6.45	6.45	0.00
30	7.2	7.4	−0.02	71	7.92	7.92	0.00
31	7.89	7.89	0.00	72	7.04	7.02	0.02
32	8.57	8.57	0.00	73	8.13	8.13	0.00
34	3.66	3.66	0.00	75	5.4	5.4	0.00
35	5.15	5.15	0.00	76	6.35	6.31	0.04
36	6.01	6.01	0.00	78	7.02	7.02	0.00
38	5.69	5.72	0.23	79	7	7	0.00
39	5.22	5.22	0.00	81	6.92	6.92	0.00

The results presented in Table 12.6 show that the results of the proposed approach GA-MLA are superior compared with those of stepwise ANN, SVM, MLR-SVM, and deep learning with LSTM-NN. The coefficient correlation is high, and GA-MLA model uses fewer molecular descriptors for predicting the *pIC50* activity.

Table 12.5 Observed and GA-SVM predicted activities for the test set

Data no.	Observed pIC50	Predicted pCI50
3	4.72	4.7
9	4.89	4.89
13	4.09	4.1
19	5.14	5.14
21	5.6	5.6
24	7.23	7.2
28	6.92	6.92
33	7.85	7.75
37	5.44	5.3
40	4.37	4.36
43	5.12	5.12
47	5.96	5.96
52	8.24	8.24
54	8.23	8.23
59	8.51	8.55
63	5.33	5.34
65	5.92	5.92
70	7.11	7.13
74	6.47	6.47
77	7.02	7.02
80	6.87	6.87

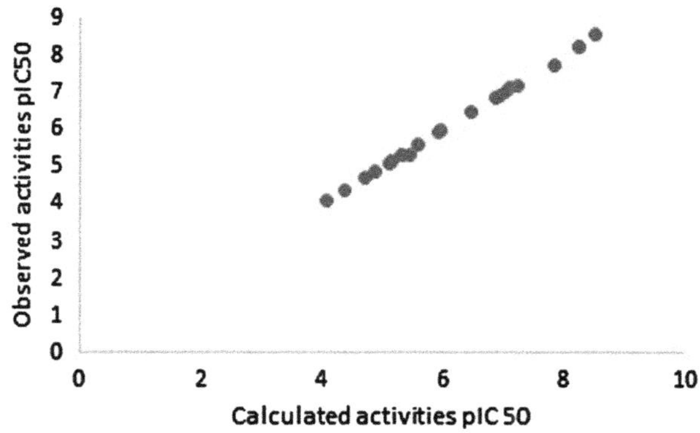

Fig. 12.3 Correlations of observed and predicted activities (data test set) values

Table 12.6 Comparison of the proposed approach with other approach

Approach	r: coefficient of correlation	Number of selected descriptors
ANN	0.92	16
SVM	0.88	16
MLR-SVM	0.89	16
LSTM-NN	0.95	16
GA-MLA	**0.98**	**3**

12.4 Conclusion

In this work, a new approach for the selection of molecular descriptors in QSAR modeling is presented. This new feature selection approach, called GA-MLA, was designed to address the problem of classification. The GA-MLA approach is based on the use of genetic algorithms which identify promising data subsets of molecular descriptors and then the confirmation of identified descriptors data subset by several machine learning algorithms.

The GA-MLA approach was illustrated and tested in a case study which constitutes an example of QSAR classification modeling, where the estimated property corresponds to the anti-HIV activity of the chemical compounds. Comparisons with the results obtained in the literature by other QSAR models were discussed, showing the potential and the usefulness of the proposed approach. The GA-MLA hybrid approach is an important idea for QSAR modeling, helping to minimize the time and money costs in the field of drug discovery.

References

1. UNAIDS, Ending the AIDS epidemic 2020: Global HIV Statistics, (2020)
2. C.P. Swathik, K.D. Jaspreet, M. Vidhi, R. Navaneethan, J. Mannu, S. Durai, Quantitative structure-activity relationship (QSAR): modeling approaches to biological applications. Encycl. Bioinforma. Comput. Biol. **2**, 661–676 (2019)
3. I. Hdoufane, J. Stoycheva, A. Tadjer, D. Villemin, M. Najdoska-Bogdanov, J. Bogdanov, D. Cherqaoui, QSAR and molecular docking studies of indole-based analogs as HIV-1 attachment inhibitors. J. Mol. Struct. **1193**, 429–443 (2019)
4. R. Todeschini, V. Consonni, *Molecular Descriptors for Chemoinformatics* (Wiley-VCH, 2009)
5. M. Eklund, U. Norinder, S. Boyer, L. Carlsson, Choosing feature selection and learning algorithms in QSAR. J. Chem. Inf. Model. **54**(3), 837–843 (2014)
6. F. Grisoni, V. Consonni, R. Todeschini, Impact of molecular descriptors on computational models, in *Computational Chemogenomics. Methods in Molecular Biology*, ed. by J. Brown, vol. 1825, (Humana Press, New York, 2018)
7. X.Y. Liu, Y. Liang, S. Wang, Z.Y. Yang, H.S. Ye, Hybrid genetic algorithm with wrapper-embedded approaches for feature selection. IEEE Access. **6**, 22863–22874 (2018)
8. B. Wutzl, K. Leibnitz, F. Rattay, M. Kronbichler, M. Murata, S.M. Golaszewski, Genetic algorithms for feature selection when classifying severe chronic disorders of consciousness. PLoS One **14**(7), 1–16 (2019)
9. K. Nagasubramanian, S. Jones, S. Sarkar, A.K. Singh, A. Singh, B. Ganapathysubramanian, Hyperspectral band selection using genetic algorithm and support vector machines for early identification of charcoal rot disease in soybean stems. Plant Methods **14**(86), 1–13 (2018)
10. H. Labjar, M. Kissi, R. Mouhibi, O. Khadir, H. Chaair, M. Zahouily, QSAR study of 1-(3, 3-diphenylpropyl)-piperidinyl amides and ureas using genetic algorithms and artificial neural networks. Int. J. Bioinforma. Res. Appl. **12**(2), 116–128 (2016)
11. N. Salari, S. Shohaimi, F. Najafi, M. Nallappan, I. Karishnarajah, A novel hybrid classification model of genetic algorithms, modified k-nearest neighbor and developed backpropagation neural network. PLoS One **9**(11), 1–50 (2014)
12. A.K. Srivastava, D. Singh, A.S. Pandey, T. Maini, A novel feature selection and short-term Price forecasting based on a decision tree (J48) model. Energies **12**, 1–17 (2019)

13. C.P. Swathik, K.D. Jaspreet, M. Vidhi, R. Navaneethan, J. Mannu, S. Durai, Quantitative structure-activity relationship (QSAR): modeling approaches to biological applications. Encycl. Bioinform. Comput. Biol. **2**, 661–676 (2019)
14. B. Liu, H. He, H. Luo, T. Zhang, J. Jiang, Artificial intelligence and big data facilitated targeted drug discovery. Stroke Vasc. Neurol. **4**, 206–213 (2019)
15. A. Racz, D. Bajusz, K. Héberger, Intercorrelation limits in molecular descriptor preselection for QSAR/QSPR. Mol. Inf. **38**, 1–6 (2019)
16. J. H. Holland. Adaptation in Natural and Artificial Systems. (Ann Arbor, MI, University of Michigan Press. 1992)
17. E. Pourbasheer, R. Aalizadeh, M.R. Ganjali, P. Norouzi, J. Shadmanesh, QSAR study of ACK1 inhibitors by genetic algorithm–multiple linear regression (GA–MLR). J. Saudi Chem. Soc. **18**, 681–688 (2014)
18. I.I. Baskin, D. Winkler, I.V. Tetko, A renaissance of neural networks in drug discovery. Expert Opin. Drug Discovery **11**, 785–795 (2016)
19. P. Pradeep, R.J. Povinelli, S. White, S.J. Merrill, An ensemble model of QSAR tools for regulatory risk assessment. J. Chem. **8**(48), 1–9 (2016)
20. T.K. Shameera Ahamed, V.K. Rajan, K. Sabira, K. Muraleedharan, QSAR classification-based virtual screening followed by molecular docking studies for identification of potential inhibitors of 5-lipoxygenase. Comput. Biol. Chem. **77**, 154–166 (2018)
21. K. Lee, M. Lee, D. Kim, Utilizing random Forest QSAR models with optimized parameters for target identification and its application to target-fishing server. BMC Bioinf. **18**, 75–86 (2017)
22. L. Wen, Q. Li, W. Li, Q. Cai, Y.M. Cai, A QSAR study based on SVM for the compound of hydroxyl benzoic esters. Bioinorg. Chem. Appl., 1–10 (2017)
23. H. Tanaka, H. Takashima, M. Ubasawa, K. Sekiya, I. Nitta, M. Baba, S. Shigata, R.T. Walker, E. De Clercq, T. Miyasaka, Structure-activity relationships of 1-[(2-hydroxyethoxy) methyl]-6-(phenylthio) thymine (HEPT) analogues: effect of substitutions at the C-6 phenyl ring and the C-5 position on anti-HIV-1 activity. J. Med. Chem. **35**, 337–345 (1992)
24. R. Garg, S.P. Gupta, H. Gao, M.S. Babu, A.K. Debnath, Comparative quantitative structure-activity relationships studies on anti-HIV drugs. Chem. Rev. **99**, 3525–3601 (1999)
25. H. Bazoui, M. Zahouily, S. Boulajaaj, S. Sebti, D. Zakarya, QSAR for anti-HIV activity of HEPT derivatives. SAR QSAR Environ. Res. **13**(6), 567–577 (2002)
26. MMP, molecular modelling pro-Demo (TM) Revision 301 demo. ChemSW Software (TM)
27. S. Anacleto de Souza, L.G. Leonardo Ferreira, S. Aldo de Oliveira, D. Adriano Andricopulo, Quantitative structure–activity relationships for structurally diverse Chemotypes having anti-Trypanosoma cruzi activity. Int. J. Mol. Sci. **20**, 1–21 (2019)
28. L. Wen, Q. Li, W. Li, Q. Cai, Y.M. Cai, A QSAR study based on SVM for the compound of hydroxyl benzoic esters. Bioinorg. Chem. Appl., 1–10 (2017)
29. S.M. Marunnan, B.P. Pulikkal, A. Jabamalairaj, S. Bandaru, M. Yadav, A. Nayarisseri, V.A. Doss, Development of MLR and SVM aided QSAR models to identify common SAR of GABA uptake herbal inhibitors used in the treatment of schizophrenia. Curr. Neuropharmacol. **15**(8), 1085–1092 (2017)
30. S.K. Chakravarti, S.R.M. Alla, Descriptor free QSAR modeling using deep learning with long short-term memory neural networks. Front. Artif. Intell. **2**(17), 1–18 (2019)

Chapter 13
Mining Electronic Health Records of Patients Using Linked Data for Ranking Diseases

Siham Eddamiri, Elmoukhtar Zemmouri, and Asmaa Benghabrit

13.1 Introduction

In traditional healthcare systems, the doctors interpret signs or symptoms of patients for the treatment of the diseases [1]. Nevertheless, many unspecific signs and symptoms can make the doctor tasks more difficult and sometimes result in undesirable errors and thus affect the quality of services. Therefore, expert systems have been developed to simulate health decisions with enhanced accuracy to solve this problem. Hence, these expert systems emulate health decision-making with improved accuracy. This involves associations between different pieces of patient information and common patterns.

The healthcare industry, therefore, is an enormous amount of data that in the twenty-first century plays an important role in healthcare. This encourages different areas like epidemiology tracking, evaluation of health outcomes, system assessment and success assessment, planning of public health, and review of policies [2]. For public-health organizations, this knowledge is even more important for diagnosing, preventing, and dealing with diseases while facing pressure to reduce costs and improve efficiency. For hospitals and in other health centers, health computerization has already been developed and will be producing huge amounts of information from medical records, patient care, and medical imaging, such as the routine use of information systems and electronic medical device.

S. Eddamiri (✉) · E. Zemmouri
Department of Mathematics and Computer Science, University Moulay Ismail, ENSAM, Meknes, Morocco
e-mail: e.zemmouri@ensam-umi.ac.ma

A. Benghabrit
Department of Computer Science, University Mohammed V, ENSMR, Rabat, Morocco
e-mail: benghabrit@enim.ac.ma

M. Ouaissa et al. (eds.), *Computational Intelligence in Recent Communication Networks*, EAI/Springer Innovations in Communication and Computing,
https://doi.org/10.1007/978-3-030-77185-0_13

In actuality, medical data from a number of sources are needed in clinical research, such as clinics, hospitals, and medical organizations [3–6]; it is important to prepare and confirm these requirements through the extraction of information from the available data. As a result, numerous aspects of observing, requesting, and analyzing health information using linked data are becoming increasingly important as an excellent technology. In many areas, especially bioinformatics, public linked medical data is already one of the greatest technological innovations in the last 10 years. The clinical analysis of medical evidence was a major challenge, although significant efforts are being made to upload medical and health information on the Web [7]. This is due to two main limitations: (1) The most typical problem of clinical information system is the limitation of the interoperability with other health applications such as Electronic Medical Records (EMRs) [7]. (2) The heterogeneity of the data models as medical data from various sources. These are the main aims of these problems, by defining systems that incorporate medical details such as hospitals, clinical sites, research institutes, and pharmaceutical companies. These are the growing rapidly fields of study like semantic web and data mining. Sensitive patient data such as population data, diagnosis, drug and radiology images, test results, doctor entries, and comments may be included in medical data.

This chapter aims mainly to develop a semantic web system for facilitating the analysis of data from patient-oriented diseases including: (1) The extraction of the entities from the medical record of patients to link them to their concepts of medical ontologies; (2) The representation of the diseases' view by regrouping the disease according to their similar symptoms; (3) The representation of the ranked disease using a patient record. Thus, in this chapter, we describe an approach using neural language models for RDF data clusters to identify diseases using the grouped symptoms by integrating medical linked data and unstructured data; (4) A calculation of the similarity between the vector representations of the patient's medical record and the disease vectors identified for ranking disease.

In this chapter, we present related structured work on linked data in medical data and study semantic knowledge that is employed in numerous data mining applications in healthcare applications. We then define our methods of labeling the disease of patients by obtaining vectors from the RDF graph so that the clustering process is ready to produce disease for the clusters within the Medical RDF dataset. The results of the analysis are then achieved. There are conclusions in the final section.

13.2 Related Work

Clinical studies seek to improve the way disease is understood, diagnosed, prevented, and treated. Medical data from several sources such as clinics, hospitals, and pharmaceutical organizations are often required to be included [3–6]. The use of clinical (unstructured data) and biomedical information is an outstanding technology to improve patient understanding and reveal biomedical relationships of interest.

The linked data defines a way to view, share, and connect data through the Web [8, 9]. The semantic web is steadily developing, organized, and interlinked in the life sciences, through the use of W3C Resource Description Framework (RDF) and OWL [10]. BioPortal has actually developed almost 400 open biomedical ontologies [11] and several biomedical datasets available on Bio2RDF [12]. For example, the authors in [13] categorize disease and other health statuses using the 11th International Classification of Diseases (ICD11). In addition, the Experimental Factor Ontology (EFO) [14] includes types of reference ontologies, such as disease, cell line, cell type, and anatomy. There has also been a substantial increase in biomedical ontology in the number of studies in connected medical data. For example, Protein sequence databases UniProt [15], Biological pathways are identified in reactome [16], ChemSpider in chemical compounds ChemSpider [17], small-molecular bioactive drug-like ChEMBL [18]. For example, the paths in biological process are identified.

Moreover, there has also been a huge interest in awareness of linked medical data in order to improve biomedical ontology. For instance, UniProt protein sequence databases [15] and biological pathways are identified in reactome [16], ChemSpider in chemical compounds [17], and pathway in WikiPathways [19], small-molecular bioactive drug-like ChEMBL [18]. However, sensitive patient information, like populations, diagnostic and medical devices, radiology images, laboratory results, medical data, and observations, can be presented in medical information [20, 21]. Clinical researchers encounter obstacles ethically [7] and legally in the discovery and accessibility of technical and interoperability medical data [22]. Thus, such ethical and legal problems need urgently to be resolved by the subjects with special characteristics not known or anonymized.

With this huge and accessible public biomedical ontology and linked medical data nowadays, there is an urgent need to be used efficiently. We must therefore gather useful different techniques and turn such information into valuable knowledge that will result to the positive development of resource use, patient health, and biomedical process improvement. However, the diagnosis of a disease in traditional medicine environments depends upon the choice of the doctor to make it the most probable cause depending on the symptoms of one person. It might also result in an accidental error leading to increased healthcare costs and affecting patient quality of treatment. Data mining has therefore recently been instrumental in forecasting safety or in diagnosing high-quality conditions. Thus, data mining offers different methodologies to discover useful knowledge [23], for example, disease detection [24], low-cost medical treatment detection [25], or even health insurance fraud detection [26]. Indeed, the next phase is to feed the learner with functional vectors by choosing the right data mining algorithm depending on the user's aims when determining the correct data, preprocessing, and extracting instances converted into vectors. Generally, machine learning algorithms could be classified into the following categories: (1) Supervised techniques of learning using predefined input–output pairs to learn the input–output mapping function

[27]. The relationship between diagnoses and conditions, genome, and illnesses predicated on conditional random fields is extracted from Bundschus et al. [28]. On the other hand, Shouman et al. [29] classify heart disease diagnosis methodology using neighbor K-nearest (K-NN). However, the authors in the work [30] predict the breast cancer and the chronic disease patients' survival. Coden et al. represent a system by which unstructured reports like drugs and diagrery classify and identify the corresponding entities [31]. Poesio et al. [32] are currently working on extracting from text a number of concept features in psychological studies, which are actually linked to concepts, using the state-of-the-art relationship extraction methods. (2) Unsupervised learning techniques rely on the knowledge of the input patterns and the determination of relationships without predetermined outputs between them [33, 34]. Legaz-García et al. [35] are developing open data mappings of biomedical systems, using aggregatory clustering, among entities of data scheme and ontological infrastructure. Escudero et al. either categorize Alzheimer's disease into pathological categories or identify breast cancer recurrence by K-means [36]. While Missikoff et al. [37] combine a language method with a statistical method for automatically generating OntoLearn ontology from text.

Methodology
In this section, we present the details of our system for integrating RDF and unstructured data to support machine learning medical decision systems; see Fig. 13.1. Our contributions can be summarized on three steps depicted as the following: (1) First, we extract the entities from the medical record of patients to link them to their concepts of medical ontologies like RadLex, FMA, SNOMED CT such as medical images, reports, and laboratory results as a training file. (2) Second, for each entity, we extract its substructures to train them using Doc2vec to generate vectors with high dimension, then, we use the K-means algorithms to regroup similar entities into homogeneous clusters (symptoms), and next, the themes (diseases) of clusters are identified. (3) Finally, we calculate the similarity between vector representations of medical data extracted from this step and diseases generated from the first step to rank likely disease.

13.2.1 Mining Electronic Health Records Using Linked Data

Medical data from various resources, such as radiology pictures, laboratory results, medication entries, and comments, are often required to be included in the clinical patient data for understanding, curing, detecting, and managing diseases. However, due to the enormous amount and complexity of these data, clinicians can discourage them all from making full use of these data. Thus, by linking the symptom ontology of the diseases with other medical ontologies with two reasons, we must first identify entities from text to specific interest concepts: This is first of all essential because we are trying to analyze annotations taken from them. Our second objective is to appreciate its structure [38]. Then, we discuss how we can vectorize medical

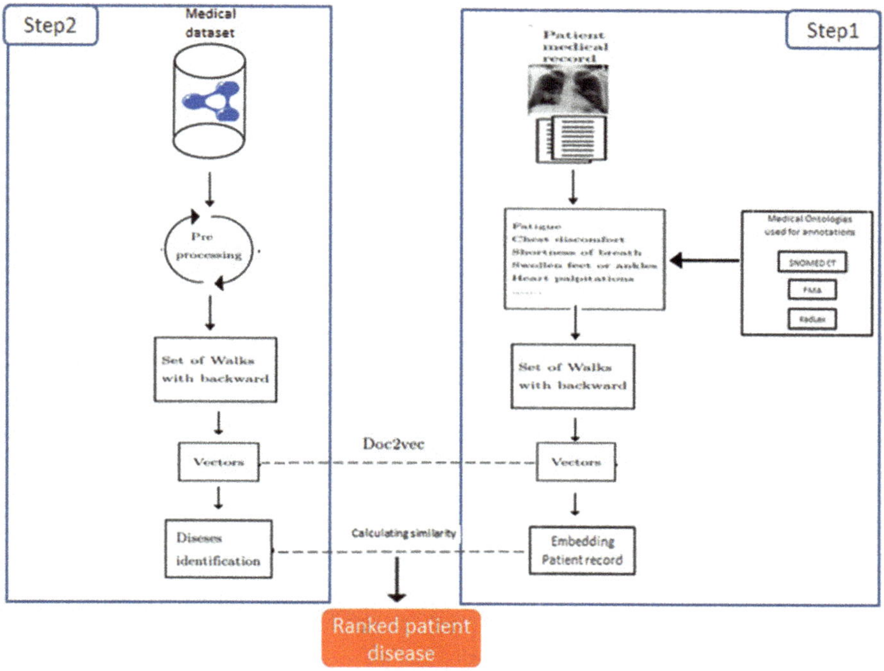

Fig. 13.1 Our methodology to support a medical decision system using a mining Electronic Health Record for Ranking Diseases

annotations by using Doc2vec and TF-IDF to delete the smallest data to every file (patient medical record) [39, 40]. Then, we utilize Doc2vec to create the numerical values in a latent space in each entity for neural network models [41]. For the trainings file, we use Doc2vec to generate a model file. A model file is produced by the Doc2vec tool. It first builds a vocabulary in the training file and then acquires parameters to produce large-size word vector representations. The Doc2vec codes texts such as sentences, paragraphs, or whole documents as vectors. Doc2vec is used to train the training file with the goal of obtaining a model file. The Doc2vec tool produces a model file and a training file. It first builds a vocabulary in the trainings file and then learns high-dimensional word vector representations by parameters.

13.2.2 Diseases Identification Based on Clustering RDF Dataset

This section aims to propose an integrated approach that identifies suitable diseases automatically for the generated medical data clusters. The following steps describe the calculation of these recommendations: (1) After the generation of the embedding vectors from the previous phase (Sect. 13.2.1), we use the K-means algorithm to extract heterogeneous entities into clusters (symptoms) of similar entities. (2) Then, we rank disease by calculating the similarity between the embedding patient's medical record vectors and diseases identification produced from the first step. This section provides a clearer and integrated approach to identify the appropriate diseases for the clusters generated by the approach. This approach is designed with the extraction of labels or tags to the RDF graph clusters to define a disease-centered label on an RDF graph (see Algorithm 1). Since medical data are from different sources, many situations can lead to information loss (entities and links), specifically when ontologies have many abstract classes that make it difficult to identify medical data diseases. We therefore intend, by generating feature vectors of entities, to apply the above described algorithm. We then develop a set of labels for each cluster to describe diseases (theme th) through the TF-IDF measure in order to determine the most common of rdf:disease. The diseases are identified by the class of disease that arises more in the cluster concerned. In the Th theme, vr is the node in a gr group. In each group, the theme annotation is associated with the most common nodes e as follows:

$$Th(cluster) = argmax_{e \in \mathcal{E}}(count(v \mid (v, dbp : Disease, e) \in \mathcal{T})),$$

where $\mathcal{E}$ represents the set of all types defined as

$$\mathcal{E} = \{e \ \in V \mid (x, dbp : Disease, e) \in \mathcal{T}\}.$$

Finally, we calculate the similarity between vector representations of medical data extracted from the second step and diseases generated from the first step to rank likely disease based on cosine similarity [34]. Thus, if the entered patient record vector is close to the specific centroid, this patient record will share semantic and syntactic contexts with a large part of the symptoms in this group; the distance is then calculated by using cosine similarity. The entered patient vector records will then be calculated with the clusters (symptoms and diseases) (disease). Then, we have ranked the disease that matches the patient's symptom information. Thereafter, we measure the vital signs of the ranked disease such as sex, body temperature, pulse rate, etc.

13.3 Results and Discussion

13.3.1 Dataset

Two datasets were used in this chapter. The first one was obtained from the BioPortal [11], and it consists of 30 diseases–symptoms graph. The second one was covid disease dataset [42].

Algorithm 1: Ranking diseases-symptoms algorithm

```
Input  : set of RDF triples, size, k
Output: ranked diseases
/* k : is cluster number depending on the dataset */
for v_r ∈ V do
    Extracting a set of walks-with-backward rooted in v_r with depth d;
    Creating Document for each extracted set of walks-with-backward;
end
for Doc ∈ Documents do
    Eliminate low frequent entities using TF − IDF;
end
Run Doc2vec;
Initialise the K-means by choosing k random centroids;
repeat
    (re)assign each instance to the closest centroid using Euclidean distance;
    Update the clusters centroids, i.e., calculate the mean value of the instances for each
    cluster;
until no change;
return Clusters;
for cluster ∈ Clusters do
    /* Extract Theme (disease) using Th */
    Th(cluster) = argmax_{e∈E}(count(v | (v, dbp : Disease, e) ∈ T))
end
for th ∈ themes do
    /* calculate similarity between the theme th and each
       embedding Patient record P using: */
    Sim(th, P) = Cosine − similarity(th, P);
end
return Rankedpatientdisease;
```

- **BioPortal dataset**. The Ontology Biomedical Repository [11] provides access for the public to over 400 ontologies and 6 million ontological entities. In the biomedical field, it appears to be the largest repository of ontologies. BioPortal's ontologies cover a range of biomedical areas, including diseases, phenotypes, clinical observations and findings, genes, proteins, etc. For our scenario, we generate 30 diseases–symptom graph from BioPortal such as heart disease, allergy disease, lung disease, etc.
- **The covid disease dataset**. This dataset [42] was created on the basis of a report published on Kaggle containing data from over 63,000 papers for the analysis of the vast mass of the literature. All data in the original dataset is linked in this knowledge graph to facilitate the search for relevant information. Additional links to relevant existing external resources will also enhance the graph of the knowledge.

13.3.2 Validation Metrics

Two tests will determine the efficiency of the clustering phase. They make it possible to compare the clusters with known groups to evaluate the clustering process reliability. First of all, the datasets of the pre-created clusters have been defined. After this, the F-measure and purity parameters are calculated. The F-measure is based on the fundamental criteria of the recovery of knowledge, "precision," and the concept of "recall":

$$P(i, j) = \frac{n_{ij}}{n_j} \tag{13.1}$$

$$R(i, j) = \frac{n_{ij}}{n_i}, \tag{13.2}$$

where n_i is the number of the vertices of the class i, n_j is the number of the vertices of the cluster j, and n_{ij} is the number of the classes i in the cluster j. The F-measure of a cluster and a class i is given by

$$F(i, j) = \frac{2P(i, j) * R(i, j)}{P(i, j) + R(i, j)}. \tag{13.3}$$

For the entire clustering result, the F-measure is computed as follows:

$$F = \sum_i \frac{n_i}{n} \max_j(F(i, j)), \tag{13.4}$$

where n is the total node number. The biggest F-measure measures a better clustering result. Second, the purity of a cluster is the ratio of the dominant group within a range to the size of the cluster. The purity of the cluster is therefore

$$purity(j) = \frac{1}{n_j} \max_i(n_{ij}). \tag{13.5}$$

The overall purity value is the balanced average of all purity values. The formula is as follows:

$$purity = \sum_j \frac{n_j}{n} purity(j). \tag{13.6}$$

The biggest purity is a better clustering result [43].

Experimental Results and Discussion

Experimental Process

The effectiveness of the clustering phase will be determined by two tests. They allow for a comparison of the clusters obtained with known groups to assess the reliability of the clustering process. First of all, the datasets of the pre-created clusters have been defined. Therefore, on two integrated datasets, we assess our clustering process and disease identification algorithm to rank patient diseases:

- First, by constructing walks-with-backward with the depth of $d = \{2, 3\}$, we transform the RDF graphs into a number of sequences. We then use those sequences to train the DBOW & DM model with the parameters below: Window size = 4; iteration numbers = 20; negative samples = 20; vector entity size = {300}.
- The K-means algorithms are then used as inputs for numbers and a number of observation vectors that we cluster them into homogeneous groups.
- After the cluster has been identified, the user will be given both the correct disease and a summary of the contents of each cluster. We use the name of the disease label, which represents the most important (frequent) form of the groups, to be extracted.
- Finally, we calculate the similarity between vector representations of medical data extracted from the second step and diseases generated from the first step to rank likely disease.

In the following sections, the results are summarized and discussed. The aim of these experiments is to compare our system with a method used to evaluate disease identification by combining the walks-with-backward and skip-gram and the combination of set of walks with TF-IDF and Doc2vec.

13.3.3 Quality of Results and Discussion

The results of the patient disease ranking procedure with the mining integration of unstructured data and linked data are studied and discussed in this section. The results for our methodology with the following combination: the combination of set of walks-with-backward with TF-IDF and Doc2ec, the combination of walks-with-backward and skip-gram, and the combination of set of walks with TF-IDF and Doc2vec are given in Figs. 13.2 and 13.3 in terms of purity and F-measure, respectively. The results show that all the combinations can rank the patient disease relatively correct. Our system is the most accurate algorithm confirmed by the results of purity and F-measure. From our analysis, the combination of set of walks-with-backward with TF-IDF and Doc2ec listed first in all evaluation, followed by the combination of set of walks with TF-IDF and Doc2ec in second position. This can give us a substantial conclusion on the role of set of walks-with-backward in sequence creation, which is more precise and gives valuable knowledge than the technique of walk and set of walks. In addition, the elimination of the instances that give little information using TF-IDF made a positive contribution to ensemble learning.

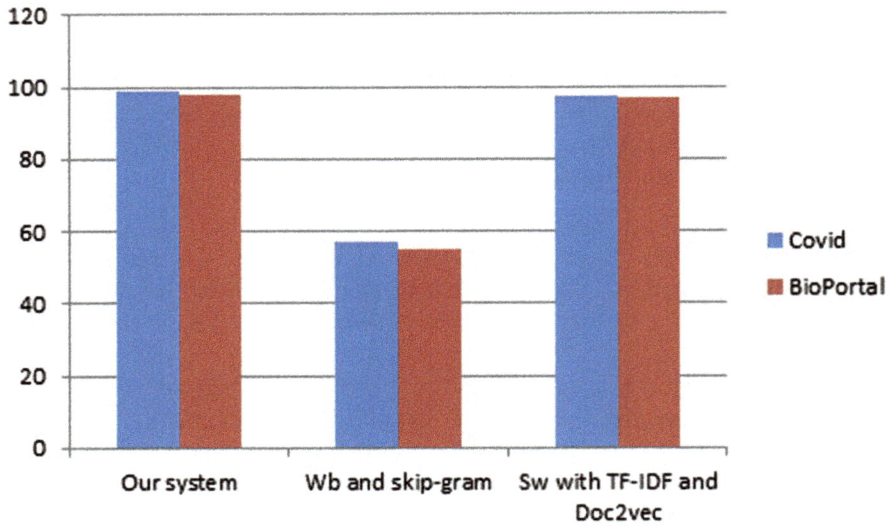

Fig. 13.2 The purity of our system and the other evaluated methods

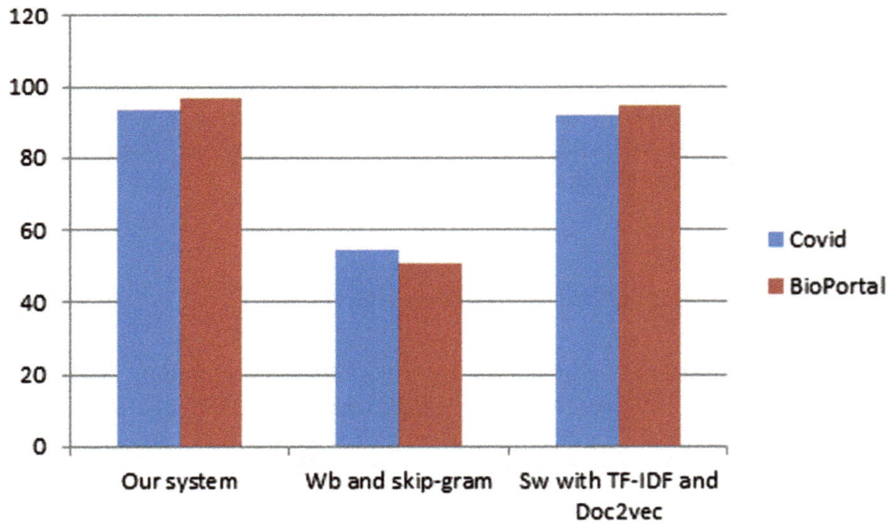

Fig. 13.3 The F-measure of our system and the other evaluated methods

13.4 Conclusion and Future Work

The innovative development of Tim Berners-Lee's World Wide Web is rapidly expanding, particularly in the health sector. Consequently, the semantic web plays an important part in information sharing, data interoperability, and information exploration. However, there exists a communication gap between the biomedical linked and unstructured data to rank patients' disease.

Therefore, we presented a clustering approach to define the suitable disease of the medical entities, so that patients' disease can be automatically ranked in this chapter. This enables us, along with the disease–symptoms proposed methodology, to evaluate and map the relations between two different data types. Our system includes the following phases: (1) The extraction of the entities from the medical record of patients to link them to their concepts of medical ontologies; (2) The generation of substructures from graph RDF using a set of walks-with-backward method; (3) The construction of embedding vectors using neural language models Doc2vec and TF-IDF as a preprocessing phase; (4) The K-means algorithm is used to regroup similar entities into homogeneous clusters (symptoms) algorithm; (5) The identification of diseases by extracting from the clustered entities the most relevant dbp:Disease; (6) The calculation of the similarity between the grouped diseases–symptoms with the unstructured medical record of the patient. Through our experiment, we have confirmed that the combination of set of walks-with-backward with TF-IDF and Doc2vec techniques gives promising results, as the clustering results leading to our system tests are over 98% in terms of F-measures.

We plan to use our methodologies in our future research to an efficient prediction method for the purpose of improving and reducing medical care. Moreover, to further improve the quality of the result, we will use a different combination of machine-learning-based classifiers and semantic reasoners. In addition, it is possible to increase the clustering algorithm. To this end, we propose comparing various mapping sources and optimizing clusters to avoid path-like clusters for each cluster.

References

1. H. Morowitz et al., *Models for Biomedical Research: A New Perspective*. (National Academy of Sciences, Washington, DC, 1985)
2. J. Studnicki, D.J. Berndt, J.W. Fisher, Using information systems for public health administration, in *Public Health Administration: Principles for Population-Based Management*, 2nd edn. (Jones and Bartlett, Sudbury, 2008), pp. 353–380
3. A. Burgun, O. Bodenreider, Accessing and integrating data and knowledge for biomedical research. Yearb. Med. Inform. **17**, 91–101 (2008)
4. M.G. Weiner, P.J. Embi, Toward reuse of clinical data for research and quality improvement: the end of the beginning? Ann. Intern. Med. **151**, 359–360 (2009)
5. J.C. Maro et al., Design of a national distributed health data network. Ann. Intern. Med. **151**, 341–344 (2009)
6. H.-U. Prokosch, T. Ganslandt, Perspectives for medical informatics. Methods Inf. Med. **48**, 38–44 (2009)
7. R.D. Kush, E. Helton, F.W. Rockhold, C.D. Hardison, Electronic health records, medical research, and the Tower of Babel. N. Engl. J. Med. **358**, 1738–1740 (2008)
8. T. Heath, C. Bizer, Linked data: Evolving the web into a global data space, in *Synthesis Lectures on the Semantic Web: Theory and Technology*, vol. 1 (2011), pp. 1–136
9. R. Cyganiak, M. Hausenblas, E. McCuirc, in *Linking Government Data* (Springer, Berlin, 2011), pp. 135–151

10. S. Eddamiri, E. Zemmouri, A. Benghabrit, RDF data clustering based on resource and predicate embeddings, in *IC3K 2018-Proceedings of the 10th International Joint Conference on Knowledge Discovery, Knowledge Engineering and Knowledge Management*, vol. 1 (2018), pp. 367–373
11. P.L. Whetzel et al., BioPortal: enhanced functionality via new Web services from the National Center for Biomedical Ontology to access and use ontologies in software applications. Nucleic Acids Res. **39**, W541–545 (2011)
12. F. Belleau, M.-A. Nolin, N. Tourigny, P. Rigault, J. Morissette, Bio2RDF: Towards a mashup to build bioinformatics knowledge systems. J. Biomed. Inform. **41**, 706–716 (2008)
13. T. Tudorache, C.I. Nyulas, N.F. Noy, M.A. Musen, *Using Semantic Web in ICD-11: Three Years Down the Road in International Semantic Web Conference* (2013), pp. 195–211
14. J. Malone et al., Modeling sample variables with an Experimental Factor Ontology. Bioinformatics **26**, 1112–1118 (2010)
15. U. Consortium, Activities at the universal protein resource (UniProt). Nucleic Acids Res.**42**, D191–D198 (2014)
16. D. Croft et al., Reactome: a database of reactions, pathways and biological processes. Nucleic Acids Res. **39**, D691–D697 (2010)
17. H.E. Pence, A. Williams, *ChemSpider: An Online Chemical Information Resource* (2010)
18. E.L. Willighagen, et al., The ChEMBL database as linked open data. J. Cheminformatics **5**, 1–12 (2013)
19. Kelder, T. et al., WikiPathways: building research communities on biological pathways. Nucleic Acids Res. **40**, D1301–D1307 (2012)
20. P.B. Jensen, L.J. Jensen, S. Brunak, Mining electronic health records: towards better research applications and clinical care. Nat. Rev. Genet. **13**, 395–405 (2012)
21. E.C. Lau, et al. Use of electronic medical records (EMR) for oncology outcomes research: assessing the comparability of EMR information to patient registry and health claims data. Clin. Epidemiol. **3**, 259 (2011)
22. P. Taylor, Personal genomes: when consent gets in the way. Nature **456**, 32–33 (2008)
23. S. Sabra, K.M. Malik, M. Alobaidi, Prediction of venous thromboembolism using semantic and sentiment analyses of clinical narratives. Comput. Biol. Med. **94**, 1–10 (2018)
24. D.S. Celermajer, C.K. Chow, E. Maijon, N.M. Anstey, K.S. Woo, Cardiovascular disease in the developing world: prevalences, patterns, and the potential of early disease detection. J. Am. Coll. Cardiol. **60**, 1207–1216 (2012)
25. D. Brixner et al., Patient support program increased medication adherence with lower total health care costs despite increased drug spending. J. Manag. Care Spec. Pharm. **25**, 770–779 (2019)
26. C. Phua, V. Lee, K. Smith, R. Gayler, A comprehensive survey of data mining-based fraud detection research. arXiv preprint arXiv:1009.6119 (2010)
27. T. Hastie, R. Tibshirani, J. Friedman, in *The Elements of Statistical Learning* (Springer, Berlin, 2009), pp. 9–41
28. M. Bundschus, M. Dejori, M. Stetter, V. Tresp, H.-P. Kriegel, Extraction of semantic biomedical relations from text using conditional random fields. BMC Bioinf. **9**, 207 (2008)
29. M. Shouman, T. Turner, R. Stocker, Applying k-nearest neighbour in diagnosing heart disease patients. Int. J. Inf. Educ. Technol. **2**, 220–223 (2012)
30. C.-L. Chang, C.-H. Chen, Applying decision tree and neural network to increase quality of dermatologic diagnosis. Expert Syst. Appl. **36**, 4035–4041 (2009)
31. A. Coden et al., Text analysis integration into a medical information retrieval system: challenges related to word sense disambiguation, in *Medinfo 2007: Proceedings of the 12th World Congress on Health (Medical) Informatics: Building Sustainable Health Systems* (2007), p. 2218
32. M. Poesio, E. Barbu, C. Giuliano, L. Romano, F.B. Kessler, Supervised relation extraction for ontology learning from text based on a cognitively plausible model of relations, in *ECAI 2008 3rd Workshop on Ontology Learning and Population* (2008), pp. 1–5

33. E. Mjolsness, D. DeCoste, Machine learning for science: state of the art and future prospects. Science **293**, 2051–2055 (2001)
34. S. Eddamiri, A. Benghabrit, et al., An improved RDF data clustering algorithm. Procedia Comput. Sci. **148**, 208–217 (2019)
35. K.-H. Cheung et al., Extending gene ontology in the context of extracellular RNA and vesicle communication. J. Biomed. Semant. **7**, 1–9 (2016)
36. J. Escudero, J.P. Zajicek, E. Ifeachor, Early detection and characterization of Alzheimer's disease in clinical scenarios using Bioprofile concepts and K means in *2011 Annual International Conference of the IEEE Engineering in Medicine and Biology Society* (2011), pp. 6470–6473
37. M. Missikoff, R. Navigli, P. Velardi, Integrated approach to web ontology learning and engineering. Computer **35**, 60–63 (2002)
38. B.J. Liu, W.W. Boonn, *Medical Imaging 2010: Advanced PACS-based Imaging Informatics and Therapeutic Applications in SPIE*, vol. 7628 (2010)
39. W. Zhang, T. Yoshida, X. Tang, A comparative study of TF* IDF, LSI and multi-words for text classification. Expert Syst. Appl. **38**, 2758–2765 (2011)
40. S. Eddamiri, E. Zemmouri, A. Benghabrit, *Theme Identification for Linked Medical Data in International Conference on Artificial Intelligence and Industrial Applications* (2020), pp. 145–157
41. Eddamiri, S., Benghabrit, A. and Zemmouri, E. (2020), "RDF graph mining for cluster-based theme identification", International Journal of Web Information Systems, Vol. 16 No. 2, pp. 223–247. https://doi.org/10.1108/IJWIS-10-2019-0048
42. B. Steenwinckel et al., in *Facilitating the Analysis of Covid-19 Literature Through a Knowledge Graph in International Semantic Web Conference* (2020), pp. 344–357
43. Y. Zhao, G. Karypis, Empirical and theoretical comparisons of selected criterion functions for document clustering. Mach. Learn. **55**, 311–331 (2004)

Chapter 14
The COVID-19 Pandemic's Impact on Stock Markets and Economy: Deep Neural Networks Driving the Alpha Factors Ranking

Badr Hirchoua, Brahim Ouhbi, and Bouchra Frikh

14.1 Introduction

The dramatic spread of COVID-19 has disrupted lives, resources, cities, and businesses worldwide. Moreover, the global measures to hold it are having a significant impact on economic activity. The COVID-19 has caused an unexpected and dramatic switch to normal modes of economic exchange and stock-market routines [31]. Until resolving the COVID-19 public health crisis, economists and investors cannot even start to predict the end of the recession that is now undertaken. Furthermore, there are many causes to expect that this downturn will be more profound and longer than the 2008 crisis. The COVID-19 pandemic represents an abnormal disruption to the global economy and trade, as production and consumption are reduced worldwide. The future economic impact of COVID-19 is highly imprecise because the spread of the virus, mortality rate, the policy responses, individual behavior, and severity are unknown [4]. Many researchers [2, 10, 32] implemented the Susceptible–Infected–Recovered (SIR) epidemiological model to analyze the economic impacts of social-distancing policies. Mckibbin et al. [27] simulated various scenarios for macroeconomic outcomes. Fornaro et al. [14] examined analytically various macroeconomic policy interventions.

The industry consequences and some international exposure impacts began to be priced in the 2–7 January period. Therefore, it is understandable that advanced investors acted on news coming out of China during this period. Moreover, starting

B. Hirchoua (✉) · B. Ouhbi
National Higher School of Arts and Crafts (ENSAM), Industrial Engineering and Productivity Department, Moulay Ismail University (UMI), Meknes, Morocco

B. Frikh
LIASSE Laboratory, National School of Applied Sciences (ENSA), Sidi Mohamed Ben Abdellah University (USMBA), Fez, Morocco

M. Ouaissa et al. (eds.), *Computational Intelligence in Recent Communication Networks*, EAI/Springer Innovations in Communication and Computing,
https://doi.org/10.1007/978-3-030-77185-0_14

from February 24, investor worries changed to corporate debt and liquidity. This decision suggests that investors understood the absolute shock and business uncertainty caused by the explosion of COVID-19 to be augmented by financial channels. The sharp drop in share prices that occurred in mid-March seems complicated to reconcile with merely cash flows being negatively affected.

The coronavirus pandemic denotes a risk that is mixing excited behavior by the global investors. In this context, Ramelli et al. [28] proposed a study demonstrating how markets are adapting to the fast rise of an earlier ignored risk. Their initial conclusions suggested that the market reasonably instantly started to react to interests in the COVID-19 possible economic consequences. Regarding future financial implications, their work examined stock price effects by capturing the expectations of market participants. They observed the stock price returns to the COVID-19 crash over three different periods. The incubation period started from 2 to January 7. Next, the outbreak period was on January 20 and February 21. Last, the fever period began on February 24. Consequently, in their second publication [29], they started with an analysis of industry-level returns over all three combined periods. Telecommunication services, food and staples retailing, and utilities performed relatively well, where energy, consumer services, and transportation were among the biggest losers. As an inauspicious sign that the crisis is potentially wide-reaching, consumer services were the biggest losers during the Fever period, and food and staples retailers were the big winners.

Baldwin et al. [3] discussed the position impacts of utilizing records and financial reasoning to supervise the forward-looking effort. The essential takeaway is that the pandemic is apt to be as contagious economically as it is medically. Al-Awadhi et al. [1] studied the influence of the coronavirus pandemic on market returns. Specifically, they applied board data analysis to examine the effect of the coronavirus pandemic on the stock markets in China. The results showed that the daily growth in total confirmed cases and death significantly affected stock returns across all companies. Gormsen et al. [15] investigated the aggregate equity market and dividend futures to quantify the shift in investors' expectations about economic growth in response to COVID-19. They concluded that the stock market fell considerably beyond changes in growth expectations due to discount rate changes. These results are consistent with Campbell et al. [8] proof on the relative importance of revisions in predicted returns in explaining aggregate market moves.

These extended explanations bring the investors to behavioral and policy reactions to the COVID-19 pandemic. COVID-19 and the containment policies have decreased directly and massively the flow of effort to businesses. The result has been an unexpected and massive discount in the output of goods and services. Voluntary choice of social-distancing practices has also operated a meaningful role. Moreover, they also have impacts that are more powerful in the modern economy for purposes sketched above. In light of this view, the policy response to the COVID-19 pandemic produces the most compelling justification for its remarkable impact on the stock market. The healthcare justification for travel confinements, social-distancing orders, and other containment strategies is clear. These policies also cause significant economic damage. New stock-market behavior is an early and noticeable reflection of the incoming damage.

Stock return predictability has been studied extensively in the literature as it reflects the economic and social organization. Significant works have been made to explain the dynamism therein. Statistics of reliable explanative power known as factors [11] have been proposed to summarize the core of predictive stock returns. Factor strategies utilize traditional investment techniques executed within a rules-based index varying from company fundamentals to security price behaviors. In the past few years, the concepts of risk factors, smart beta, or factors are common and frequent in the financial community. Factor investing is a quantitative investing methodology for portfolio creation. It contains different characteristics and behaviors obtained from varied sources, such as market data, essential information from financial statements, and sentiment analysis from the Internet.

In this context, the cross-section of stock price reactions gives a valuable chance to examine which firms' investors are mainly concerned by the direct and indirect effects of the virus outbreak and which firms may even stand to avail. Therefore, the stock market gives a continuously updated and incentivized summary of what investors believe the virus's economic consequences may eventually be. This setting also presents an unusually high-stakes possibility to study information processing in markets and how investors value various types of firm characteristics in a fast-evolving and highly uncertain context.

The Fama–French three-factor model [11, 12] is one of the most known works in this field. Malkiel et al. [26] demonstrated that multiple factors jointly determined the stock price, where the single-factor result was insufficient to define the intrinsic value accurately. Recently, Su et al. [33] created the multi-factor stock selection model by discarding repetitive factors using a fuzzy clustering algorithm. Although machine learning techniques are prevalent in stock return prediction [30], an inference of the stock returns is challenging. Nevertheless, most investors rely on their intuition to build better decision-making. Vapnik et al. [35] used the support vector machine algorithm, which became the most adopted machine learning model in the financial industry.

To help investors explore the COVID-19 pandemic impact further, we analyze the pandemic effects from a stock-market perspective. However, various intelligent approaches can be used [19, 20]. Therefore, in this chapter, we have developed an intelligent multi-factor trading strategy based on spectral clustering [36] algorithm for stock ranking that proves its efficiency over 11 years (2010–2019). Various factors, including sentiment analysis, are ranked intelligently to verify the equities returns. Besides, we utilize an LSTM autoencoder to model the market positions over time, which inputs the four latest factors and outputs their dense representation. Last, a single-layer neural network is used to transform this representation alongside the formulated ranks and the associated information coefficient to a delay period. The delay represents the period by which these ranks are applicable and profitable. The developed approach's primary purpose is to uncover multiple factors' predictive power in a single action. It attempts to find the appropriate combination of numerous alpha factors and reduce correlated/collinear factors. The developed strategy shows its extensibility over different historical event dates that significantly impacted markets. Overall, the results imply that the health crisis morphed into an economic crisis that increased through financial markets.

The rest of this chapter is organized as follows. Section 14.2 provides theoretical backgrounds in terms of investment factors and cross-sectional investment. In Sect. 14.3, we provide the developed cross-sectional investment based on clustering and intelligent delay. We then describe the pandemic's impact on the economy and markets in Sect. 14.4. Discussions are provided in Sect. 14.5. Finally, Sect. 14.6 concludes the study.

14.2 Theoretical Background

This section introduces the factors investment and the cross-sectional investment strategy by which the pandemic's impact will be analyzed.

There are many factors and investment styles in the world today. Hence, the most known strategies based on price prediction need historical data, including fundamental and pricing data. However, in the current situation, the stock markets present pending cases involving accidental changes in prices and volumes due to the COVID-19 shock. Building portfolios based on alpha factors allows investors to monitor and analyze the source and consistency of their returns more carefully. In other words, investors can compare various factors' performances before and after the COVID-19 pandemic. Companies' associated factors can provide a summarized experience about diverse companies in each sector, offering a significant picture of different market states over the COVID-19 period.

14.2.1 Investment Factors

The increasing adoption of factor investing served the fast growth of factor timing approaches for the effective allocation of multi-factor portfolios [7, 23]. Factors defining factor investment foundation are persistent drivers of asset returns restricted in the capital asset pricing and multi-factor models [13]. Regularly, various indexes are used to catch active decisions on strategic angles relative to the market. The complete adoption, lowering the price, and increasing factor indexes are extending investment strategies toward multi-factor portfolios that had better reply to strategic objectives and constraints. Besides, factors are time-dependent variables that illustrate individual factor periodicity, distinctive cumulative return patterns, and spread duration of underperformance. Hence, the progressive nature of factors provided the potential and difficulty for improving multi-factor strategies based on timing factor predictors [6, 22].

In recent portfolio theory, agents attempt to model the risk and return. However, building an optimal portfolio is highly subjective due to the failure to quantify the essential measures related to return and risk. Besides, it does not present any insight that investors can use to model their expectations. Factor attribution enables investors to identify the purpose and impact of over or underperformance by disaggregating between factor and portfolio allocation, ranking scheme, and security adoption. Factors are persistent operators of asset returns produced by various companies for providing factor indexes that improve profits, minimize risk, and enhance diversification.

There is an expanding amount of essential stock trading strategies, including cross-sectional analysis [34]. This strategy aims to construct an investment portfolio containing a large bucket of stocks applied to a solid quantitative investment strategy [16]. The strengthen point in a cross-sectional analysis lies in discovering factors with strong predictive capacities to the expected return of a trading strategy. Fama et al. [12] demonstrated that three factors could describe the cross-sectional structure of a stock price. These factors are the beta (market portfolio), the size (market capitalization), and the value (Book value to price ratio). This proof motivates numerous researchers to propose more advanced versions of factors. Harvey et al. [18] explained that the number of announced factors exhibits a rapid increase in the last two decades. Even though several single factors present a strong association with the investment strategy, the factor combinations show more robust performances. Practically, multiple factors are combined via a weighted sum, where the weights of each factor are computed by resolving an optimization with subjectively determined purpose [5]. Recently, non-linear methods such as support vector machine, random forest, logistic regression, and deep learning methodologies are applied in financial time-series modeling, yet most of them focus on the stock price prediction. In the context of cross-section modeling and feature integration, only some works exist [17, 25].

Practically, there is no unique profitability pattern or ratio, which can explain factors' profitability. The profitability ratios provide an aggregated description of business procedures, such as net profit margins and the return rate on investments. Typically, there are two kinds of profitability criteria. The first is the profitability from investments such as return on equity or the capital used, among other variations of these ratios. The second is adjusted on profit margins such as gross profit, operating profit, operating cash flow, and net profit margin.

14.2.2 *Cross-Sectional Investment*

From the expert investors' perspective, factor investing recommends a ranking-based approach by selecting securities that present distinct features based on robust and objective grounds drawn from quantitative data and applied to adopt a precise process. Frequently, stock picking deliberately focuses on the most undervalued securities, while a factor strategy maintains broad diversification over securities to decrease security-specific risk.

The cross-sectional strategy is separated into six principal stages. Success at each stage is a fundamental condition of progress; earning at a single level is not sufficient. Sufficiency stands by producing careful and sensitive decisions at each step of the process. Figure 14.1 presents the loop map of this process:

1. Universe Definition: It describes the universe of actionable elements. It should be broad with a degree of self-similarity to obtain relative value.
2. Single Alpha Factor Modeling: It defines and evaluates individual alphas used to rank the cross-section of assets in the established universe.
3. Alpha Combination: It merges various single alphas into a decisive one. It possesses more valuable forecasting strength than the best single alpha.
4. Risk Model: It prepares and determines the collection of risk factors desired to constrain a portfolio.
5. Portfolio Construction: Based on the combined alpha and the risk model, it produces a portfolio that decreases the risk under the established model.
6. Execution: It executes the trading method to transform the current portfolio into the target portfolio.

The cross-sectional equity investment strategy begins by collecting the fundamental and pricing data. Then, the investors determine the alpha signals, which need to be ranked from various factors. Therefore, for each stock, the alpha rank tries to be highest/lowest to correspond with the highest/lowest future expected return. Also, the investor will go long/short in the securities with the highest/lowest alpha-factor rank.

The next step attempts to define the objective functions, such as minimizing variance or maximizing return. Next, in the light of these restrictions, the investors introduce a risk model, such as the maximum amount of volatility allowed. Associating these signals with the determined objective functions and constraints produces the desired portfolio. More formally, these factors provide an executable trading list. It transforms the old portfolio into a new ideal one.

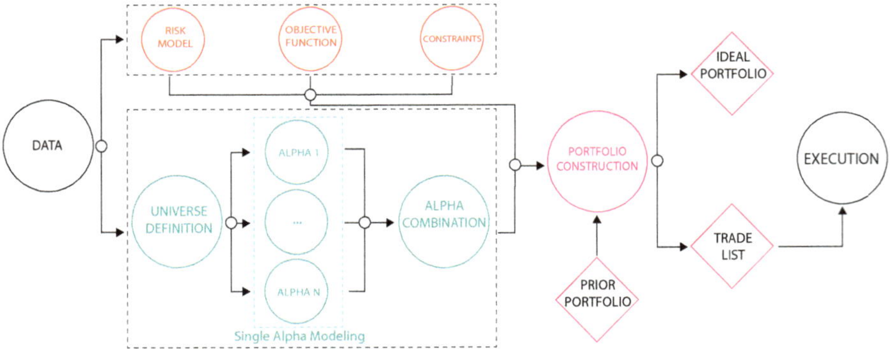

Fig. 14.1 The general strategy work flow

In Sect. 14.3, the developed strategy is further introduced, and the full process is also explained.

14.3 Cross-Sectional Investment-Based Clustering and Intelligent Delay

In this section, the developed cross-sectional investment is proposed. Precisely, the intelligent ranking-based clustering is shown, and the adaptative delay period is also detailed.

The examination of cross-sectional equity methods is motivated by the intelligent techniques-based ranking schema. In particular, this approach combines various alpha factors based on the spectral clustering algorithm. The objective is to find the predictive capability of multiple factors in one step while rejecting correlated factors. The proposed cross-sectional strategy is split into six significant steps, as shown in Fig. 14.2. First, the data modeling stage includes the data acquisition and preparation step. The system's inputs are a combination of three various sources of data that can affect future returns. Precisely, the system retrieves the pricing data of targeted stocks, collects the fundamental data of multiple firms, and selects the sentiment data used as an additional intrinsic curiosity factor. In the next step, the system preprocesses and combines the modeled data. The investor

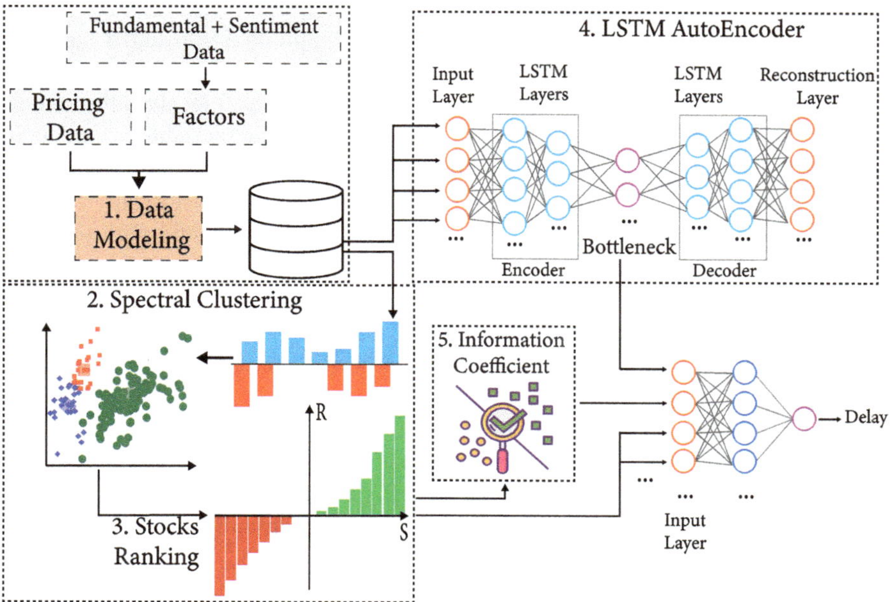

Fig. 14.2 The architecture of the developed system

determines the alpha signals needed to be ranked, which serve as features for the next steps. Precisely, transforming various factors into features and feeding them into a machine learning algorithm [9, 21]. Next, the data modeling analysis allows the investor to prune the features in the noisy financial environment. The investor starts studying the identified features. Precisely, are they consistently significant, or only in some precise conditions? What triggers a shift in importance over time? Are those essential features also relevant to other financial instruments? Are they applicable to different asset classes? What are the most pertinent features across all economic states? Pruning the feature space is an essential part of optimizing the desired models for performance and risk of overfitting. Therefore, the spectral clustering algorithm is used to rank different stocks, as the following: Given a combined dataset containing N factors, the system first builds a clustering model of k clusters. If a cluster C contains M alpha factors F_i, then each factor in this cluster is weighted as $w_i = \frac{1}{M}$. Lastly, the N factors are combined in a simple weighted sum, as follows:

$$Alpha_{combined} = \sum_{i=1}^{N} w_i * F_i. \tag{14.1}$$

We set the number of clusters to 8, which is the number of factors types used in this work, as shown in Table 14.3. For every stock in the defined universe, the alpha rank attempts to be highest/lowest to correspond with the highest/lowest future expected return. Finally, based on the learned model, the investor will go long in the securities with the highest alpha-factor rank and short on the opposite stock ranks.

The core of any algorithm is its predictive ability, predicting whether an event will happen in the future. However, investors need to measure the predictive power and be confident that they will predict future events. Therefore, the information coefficient (IC) is one of the standard measures of predictive ability, which is measured historically on data that has already been collected, reflecting how well the prediction has lined up with actual outcomes. On average, historically looking at all predictions each day, how well those predictions correspond with a result. Most canonically, the returns of individual assets are used to calculate the IC. Take the weight, which the portfolio strategy would have put on investment, and look at how that correlates with future returns. The high/low weight must connect with strong/weak future returns. On the other hand, the IC is used to testify whether or not the ranked stocks are predictive.

At each round t, an estimated ranking is built based on the developed mechanism discussed above. However, the ranked information coefficient (rank IC), which is also referred to as Spearman's correlation coefficient, is widely used in the field of finance [16] to compare the estimated rank $\hat{o}_t$ with the ground-truth ranking o_t on the universe U_t using the following formula:

$$Rank\ IC(o_t, \hat{o}_t) = 1 - \frac{6\sum_{i \in U_t}(o_{i,t} - \hat{o}_{i,t})^2}{|U_t|(|U_t|^2 - 1)}. \tag{14.2}$$

The larger the value of the rank IC is, the better a portfolio strategy based on the ranking is.

The developed approach demonstrates its strength to model and control diverse situations, including crises. However, to prolong the integration of an extensible intelligent delay definition mechanism, we introduce two other deep steps. Given the current stock's factors capable of modeling its performance, we still need to model its evolutive behavior. In other words, even though the training mechanism is robust in representing the current stock state, it lacks in predicting the time frame from which this rank is still applicable.

Therefore, in order to ensure this style, we utilize a deep LSTM autoencoder neural network. The input consists of the latest four-factor entries for each stock alongside its associated rank. It outputs a dense representation with an improved reproduction of the current state. Besides, we include the latest four stock state representations as a state history. This representation acts as an attention mechanism, where the network can focus on factors' evolution to produce a dense latent representation. The LSTM encoder layers consist of 64, 32 units, respectively, with the ReLU activation function. The bottleneck layer has 3 units. The neural network uses Adam optimizer [24] and the mean square error loss function. Next, the delay prediction neural network contains a single-layer, 68-unit that inputs the resulted representation from the last step and the information coefficient. The output is a delay period by which the current rank is applicable. Precisely, this neural network maps the input state representation and the latest information coefficient to a delay period.

The delay definition reveals the intelligent behavior by which the system controls the predictive power. This mechanism is the main focus of adjusting stocks' rank dilemma. However, each scheme is predictive of returns over a slightly different time frame. A value-based factor model may be predictive over several months, while a price-based mean reversion may be predictive over a few days.

Before proceeding to the profitability checking, the delay is observed and updated. Next, the algorithm keeps using the same value of the delay if the profitability output is checked positive. If the profitability is negative, then the delay value is redefined, the ranking values are recalculated, and the weighting scores are adjusted. If the profitability is checked negative while the delay is in the defined norms, then the factors weighting scores are revised to avoid a market crisis or unusual events. Precisely, all weights are restored to the new ranking except the influential factors that are stable in their earlier cluster. These strong factors increment their ranks by adding the weight to the last one, where the unstable factors restore their weights to the new scores only.

The developed approach has been tested in a list of historical event dates that may have had a significant impact on markets. Precisely, we have considered the US downgrade and European Debt Crisis 2011, flash crash, October 2014, market downturn (fall) in August/Sept 2015, and market regimes in terms of recovery and new normal. As shown in Fig. 14.3, this approach generates a positive profit in all events.

14.4 The COVID-19 Influence on the Markets and Economy

In order to further demonstrate the COVID-19 influence on the markets and economy, this chapter aimed to investigate COVID-19 growth analysis. Then, a comparison with the 2008 crisis is drawn. Next, the influence of COVID-19 confirmed cases on companies according to their geographic revenue is discussed. Last, based on the developed approach, which displays its performances in diverse critical situations, as shown in Fig. 14.3, the profitability delay is discussed.

The information about the pandemic is more abundant and diffused much more quickly now. Accordingly, the stock-market shock of the COVID-19 pandemic is more temporally intensive and more likely to trigger daily stock-market drops and high stock-market volatility than Spanish Flu spreads did a century earlier.

The interconnectedness of the modern economy has progressively increased the shock of the current pandemic. Precisely, the general characteristics of long-distance travel and cross-border commuting; decades of reducing communication costs, dropping transport prices, and, until lately, lowering charges; geographically dense and expansive supply chains; and the totality of real-time inventory policies are extremely weak to supply interruptions. Furthermore, the economy's composition has turned to services, many of which demand face-to-face interactions. Direct uptake of voluntary and compulsory social-distancing applications produces a sharp fall in demand for such services.

The graph shown in Fig. 14.4 reflects the interconnectedness of specific areas based on the increasing total cases. However, particular countries' disease prevention may affect the delay of investigation as well as user behavior. Precisely, it maps the velocity (speed) of the pandemic within the area. Since the virus first started in Asia, China has equally transferred the virus to other continents. Moreover, Europe

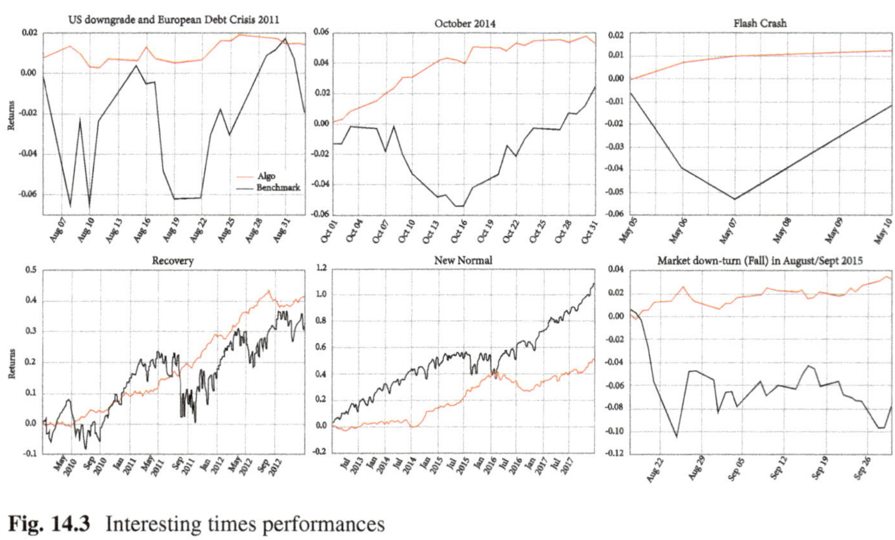

Fig. 14.3 Interesting times performances

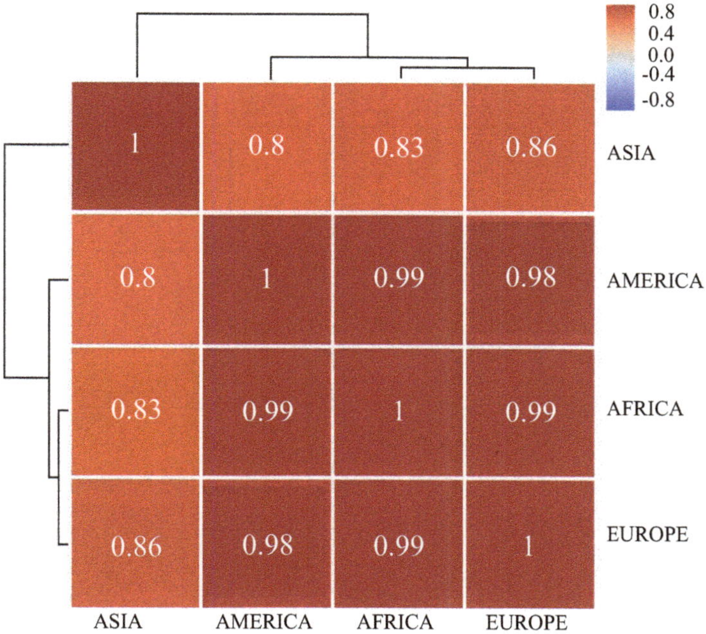

Fig. 14.4 Velocity (speed) of the pandemic over different continents

has contaminated Africa, which is clear from the same figure. Finally, America has been contaminated by different continents.

The dendrogram described in Fig. 14.5 reflects the degree of correlation between reported cases in different regions. It appears that some relationship exists in the velocity of the spread of the COVID-19 in some countries. These relationships map the pandemic behavior in each country that is influenced by some factors such as social distancing, sanitary control, and prevention. Figure 14.6 clarifies this clustering by representing five clusters of different countries.

Analysis of Minute Bar Trading Volumes of the ETFs SPY and SHY reveals that their related trading volumes could have a connection that could be employed in a trading strategy. Note that SPY is an S&P 500 ETF, and SH is a -1X (inverse) S&P 500 ETF; thus, both ETFs track a common index. Based on a 1-year baseline window for the mean and standard deviation and a 1-day window for the dollar volume smoothing, Fig. 14.7 presents the analysis of the difference in the z-score normalized trading volumes over a rolling 390-min trailing window. Moreover, this visualization illustrates a means to recognize patterns and anomalies of potential utility in implementing a trading strategy. SPY and SHY z-score-normalized minute dollar volume difference indicates a COVID-19 pandemic associated shift in the market regime, comparable to the first period of 2020 when the pandemic's economic impact had not yet been factored into the market. These results also imply

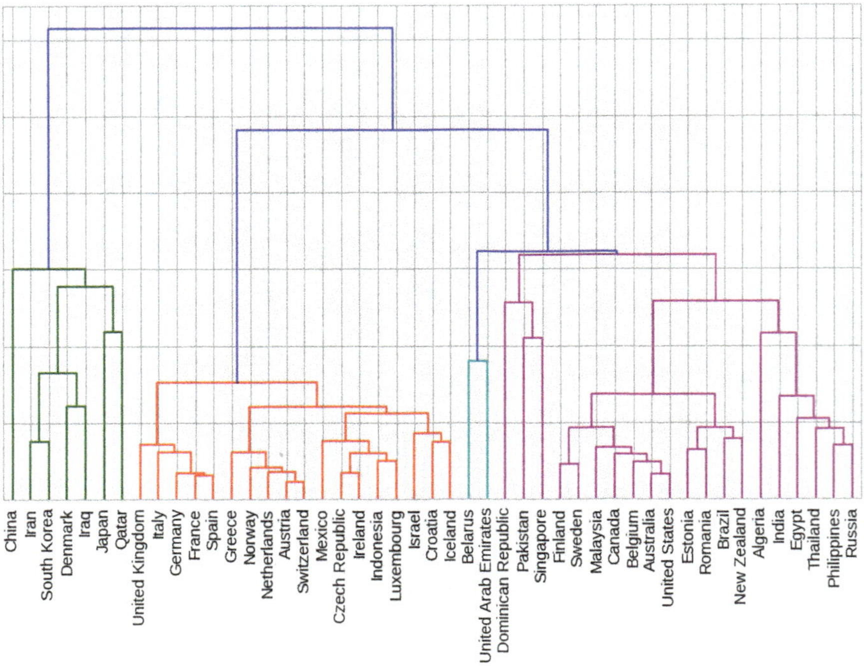

Fig. 14.5 The correlation degree between reported cases in different regions

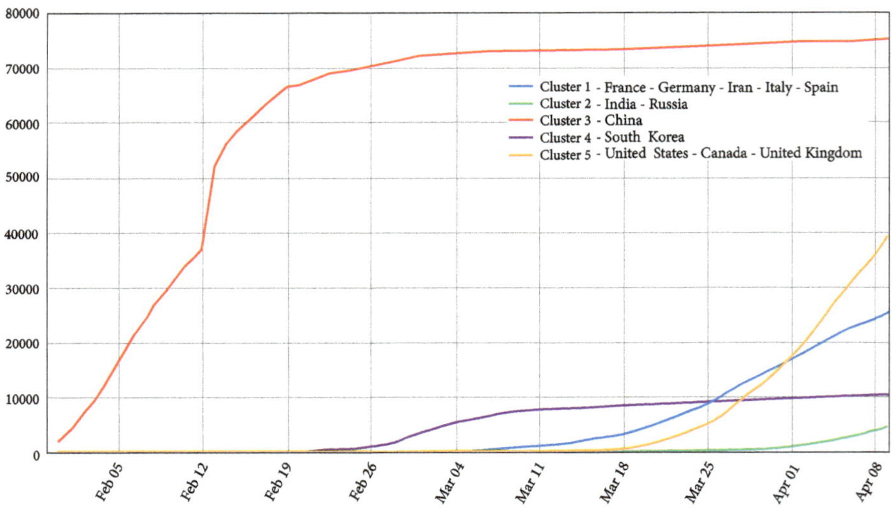

Fig. 14.6 Diverse clusters evolution according to the number of confirmed cases

that United States (US) of America's government market interventions may play a role in forcing changes in market management.

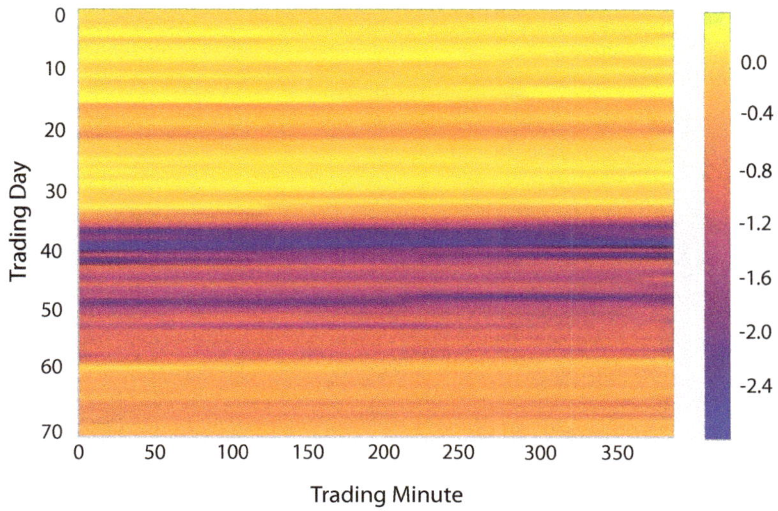

Fig. 14.7 SPY and SHY z-score-normalized minute dollar volume difference. The horizontal and vertical axes are daily trading minute and trading day, respectively; each row represents a trading day and each column a trading minute (From "2019-01-01" to "2020-04-10")

14.4.1 COVID-19 Growth Analysis

Figure 14.8 outlines the CAPM-adjusted returns for a chosen subset of crucial industries. Interestingly starting from January 21 period, US firms in Healthcare and Utilities (Energy and Transportation) start experiencing the complete loss, where approximately one-third occurred already in the precedent period. Besides, strikingly, in the last week of February, there was a briefly sharp reversal in the relative returns (as investors could sell all their stocks), followed by a whipsaw pattern (as in the aggregate market).

COVID-19 movement restrictions and social distancing have severely influenced some stocks and sectors. The following analyses seem very plausible that the growth rates of COVID-19 would affect market movements. In order to maintain the heaviest changes in various industries and sectors, Fig. 14.9 plots 1–4, 10, and 15-day future returns' correlations with various sectors and with different stocks. There is a negative correlation between the US cases proportion of the world and the returns, which indicates a potential alpha source. The Energy sector was struck with the same US proportion factor. The industries that have been hardest hit include tourism and leisure, precisely airline companies, retailers (except food and drug retailers), and some large manufacturing industries. Some industries have defeated the market at the other end of the spectrum, including food and drug companies and retailers, utilities, high-tech manufacturing, and tobacco. Unsurprisingly, as presented in Fig. 14.10, healthcare firms and biotech research have also defeated the market by a 16% falling relative to the overall fall of 35%. Moreover, the

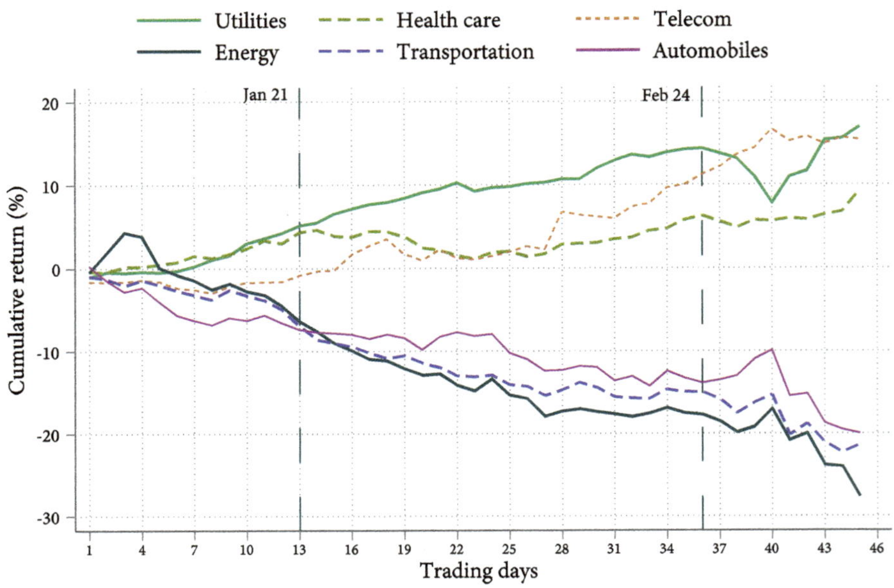

Fig. 14.8 Relative winning and losing US industries (From "2020-01-02" to "2020-03-06")

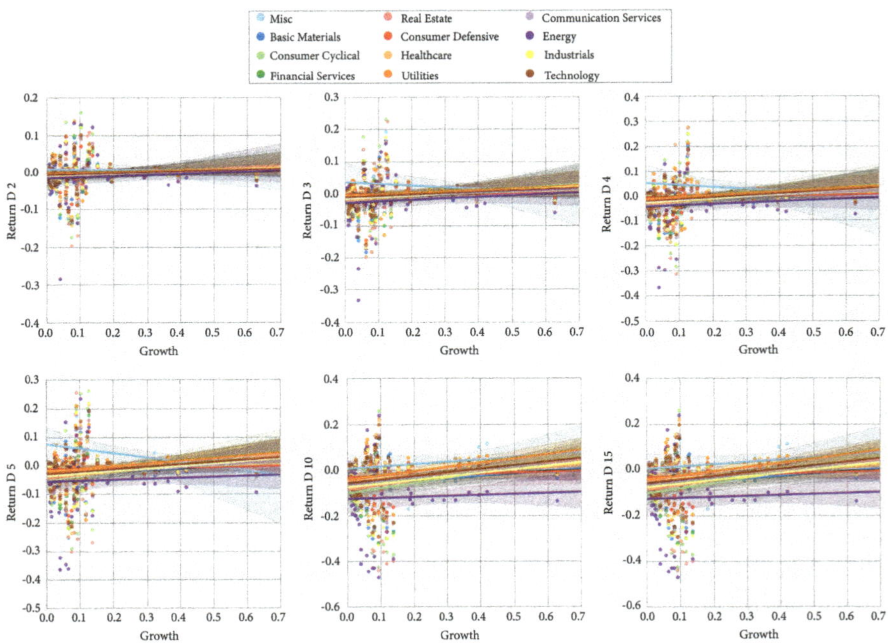

Fig. 14.9 Relative winning and losing US sectors (From "2020-01-02" to "2020-03-06")

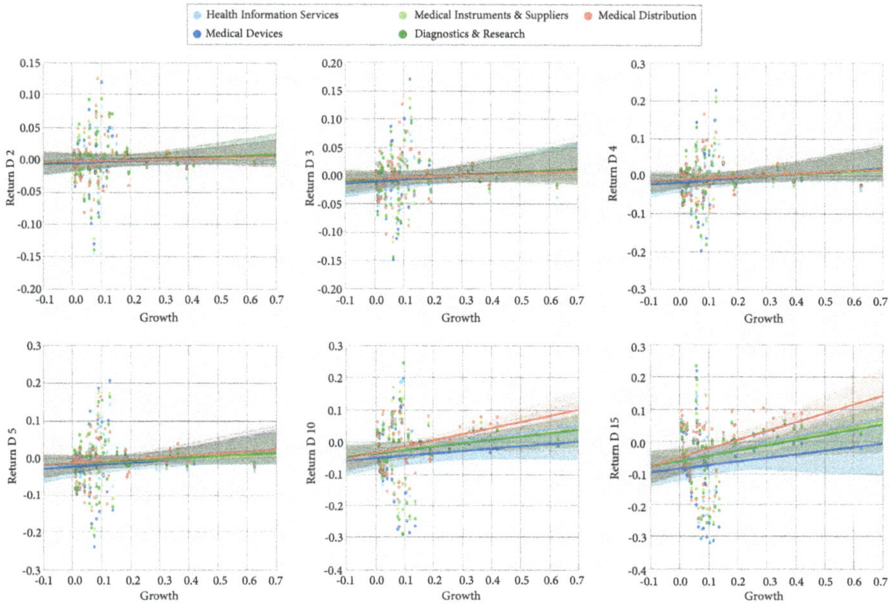

Fig. 14.10 Healthcare growth analysis (From "2020-01-02" to "2020-03-06")

insurance sector has been active in the disease period, as presented in Fig. 14.11. Finally, IT stocks are also awaited to be less hurt due to their capacity to function remotely. However, reduced-order flows due to customers constricting their spends could damage their bottom lines this financial year.

14.4.2 2008 Crisis Comparison

Valuations of several huge capitalization businesses have gone down to desirable levels, considerably below their 3-year average price-earning multiples. Notwithstanding, the earnings and revenue growth factors count on near-term attraction in stock returns. On the one hand, in the June quarter, they will grow for the most prominent businesses, and the general growth for firms is additionally expected to be highly lower for financial year 21 (FY21). Many investigation articles denote a downward review of approximately 20% in FY21 profits due to COVID-19. On the other hand, the stock prices' drop has been driven by external portfolio investors. They are drawing capital out of all emerging exchanges due to rising risk aversion. These risks will not decrease until the virus is contained.

Financial markets that continued hopefully sailing on favorable winds of liquidity have been suddenly crashed by an enormous selling wave that has caused price hitting. Until mid-February, investors were unconcerned that the coronavirus would

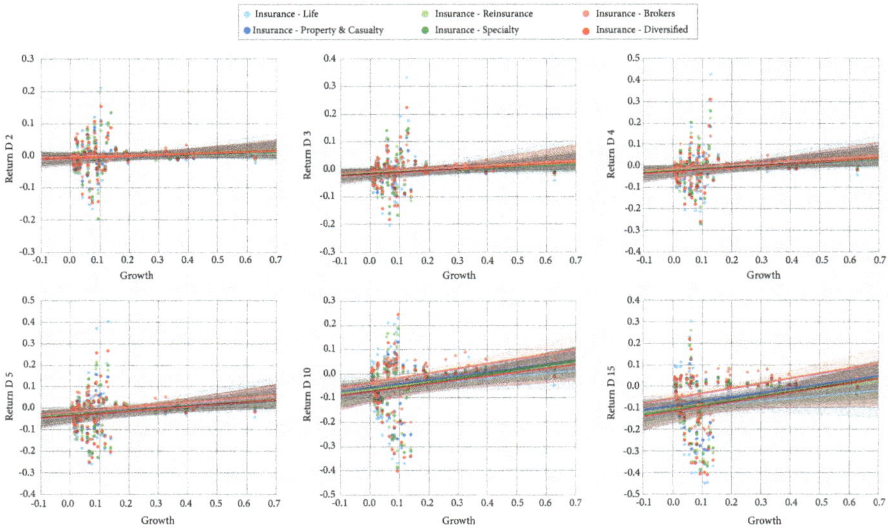

Fig. 14.11 Insurance growth analysis (From "2020-01-02" to "2020-03-06")

Table 14.1 Most serious drawdowns from 2008 to 2020. The bold values represent the maximum Net drawdown

Net drawdown %	Peak date	Valley date	Recovery date	Duration
17.28	2008-05-20	2009-03-03	2009-11-25	397
14.04	**2020-03-06**	**2020-03-18**	NaT	NaN
8.96	2015-02-02	2016-01-11	2016-06-03	350
7.17	2016-07-08	2016-12-01	2017-08-03	280
6.96	2018-08-29	2018-12-24	2019-02-19	125

be carried inside China and few countries. Unfortunately, the virus was set to wreak destruction through Europe and the USA that panic seized investors. The S&P 500 lost 33% in only 20 sessions, one of the fastest drops in recent times. The CBOE VIX, the investor doubt gauge, hit 85.4, a level last reported in 2008, and the CRB index that tracks commodity prices is at its lowest in two decades. The COVID-19 crisis threatens the entire society, making this market fall different from 2008, 2001, or 1992 crashes. Moreover, it had erupted when the global economy, particularly the US economy, was on the right sound footing, and stock prices were high.

Table 14.1 illustrates the worst drawdown periods dating from 2008 to 2020. The highest drawdown is dated to 2008-05-20, which has recovered in 2009-03-03. However, in the current pandemic context, the worst drawdown is dated to 2020-03-06. This means that the ongoing shock could be the worst one in decades. In order to demonstrate that the current situation could be critical, Table 14.2 shows the worst drawdown only for this pandemic period. Only in 4 months, the market has fallen five times, including the highest shock discussed above.

Table 14.2 Most serious drawdowns in 2020

Net drawdown %	Peak date	Valley date	Recovery date	Duration
14.04	2020-03-06	2020-03-18	NaT	NaN
2.56	2020-02-21	2020-02-27	2020-03-05	10
2.00	2019-10-04	2019-11-11	2019-11-25	37
1.15	2019-11-26	2019-12-02	2019-12-26	23
0.65	2020-01-03	2020-01-08	2020-01-14	8

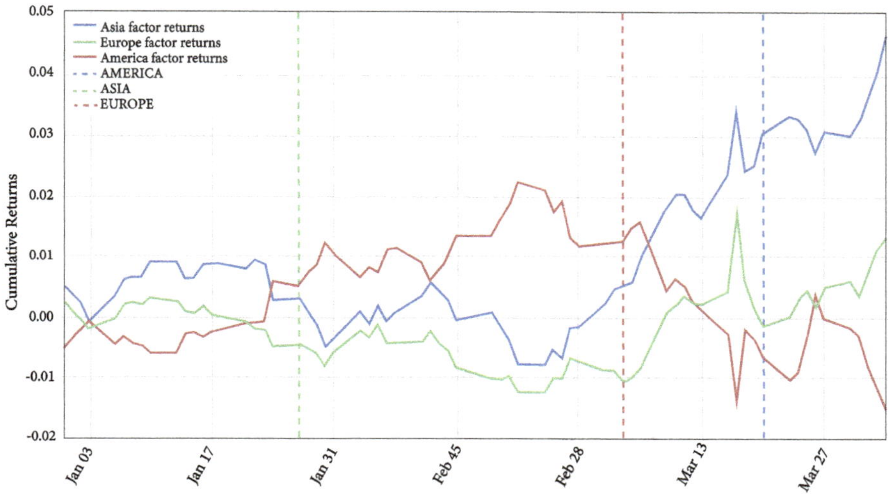

Fig. 14.12 Geographic revenue weighted factor returns in response to COVID-19 cases crossing 10,000 cases

14.4.3 The Influence of COVID-19 on Companies According to their Geographic Revenue

The confirmed COVID-19 cases impact companies according to their geographic revenue. Precisely, the following analysis loads the country-level COVID-19 confirmed cases and deaths based on their geographic revenue exposure, which validates how much income these companies are making from different regions. Figure 14.12 presents the cumulative returns over different studied areas and shows how different companies change their performance over time as a new situation is acquired. The vertical lines represent the days where each region has crossed the threshold of 10,000 confirmed cases. Figure 14.13 demonstrates these findings by providing more details. It illustrates the impacts of COVID-19 on the economy and the stock market from different regions' perspectives. Precisely, it confirms the factor returns, factor return volatility, and factor spread. However, the factor returns seem to hold well, even in the stress period, as they slowly came back to normal. The standard deviation of factor returns does not seem to be affected by the reported

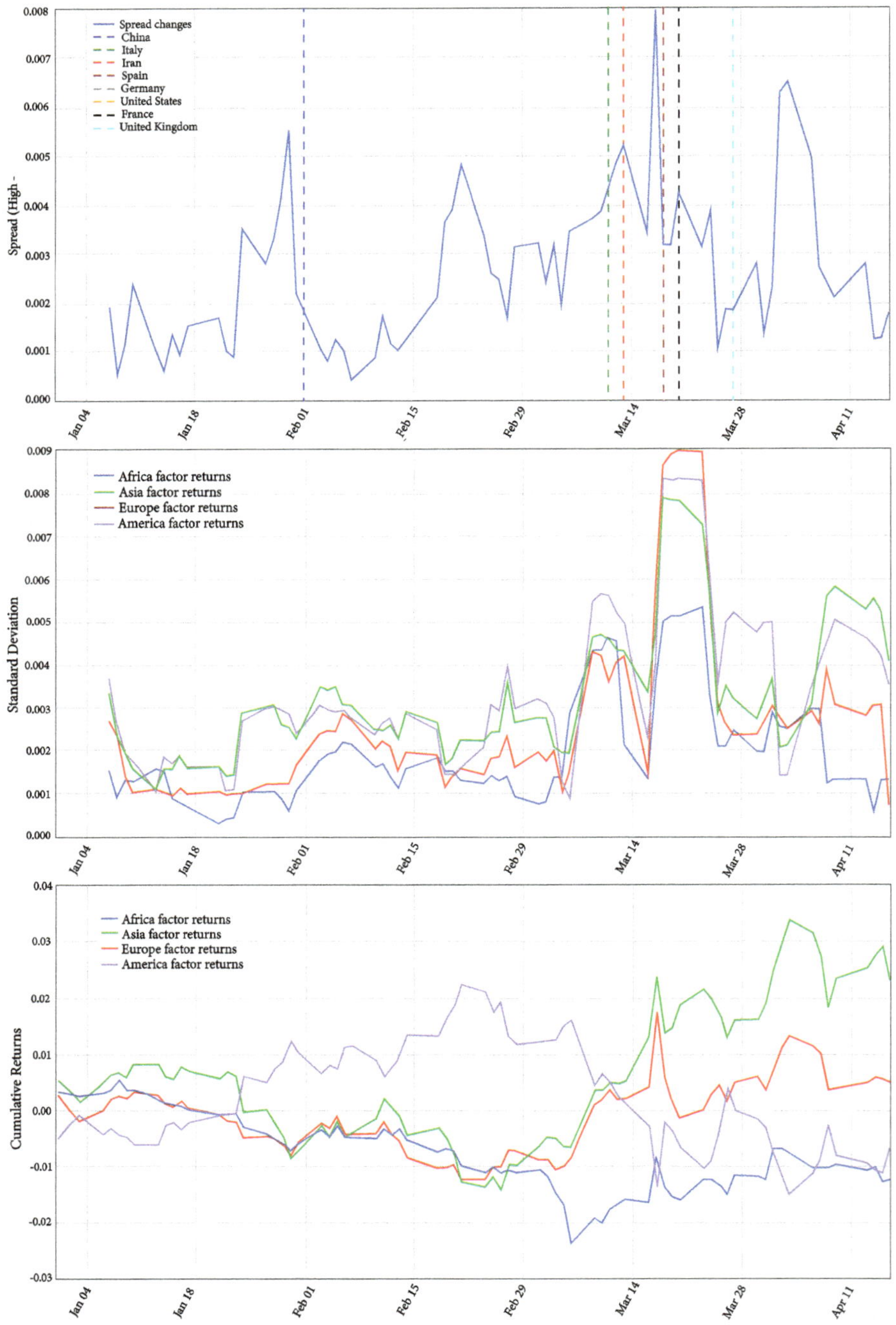

Fig. 14.13 Visualization for mapping related areas. The first figure represents the geographic revenue weighted factor returns in response to COVID-19 cases. The second figure illustrates the geographic revenue weighted weekly standard deviation of mean factor returns in response to COVID-19. The last figure shows the weekly spread of factor returns l bars countries crossing 10,000 cases

cases until cases in Europe and the USA were rising. The spread of factor returns indicates a higher spread than the average level.

14.4.4 Delay

There are many investment strategies and systems in the world today. However, each system is predictive of returns over a slightly different time frame. It is essential to define the delay over which the model is predictive and statistically verify before executing the trading strategy.

For some portfolios, investors might want to be more active in a trading strategy because they found some factors that actively have been predictive in the past. For example, if a particular signal is only predictive over shorter time horizons, why considering its old positions that do not have predictive power anymore.

The current day's factor is not tradeable on the same day, but at least the next day. However, it could be more delay in the trading strategy due to some restrictions such as turnover or risk restrictions. For example, a factor that turns over 20%, but the portfolio only becomes over 10% on that particular day, the investor only trades into virtually half of the desired portfolio. Then, the next day the investor can trade the other 10% and unless the portfolio moved again. The higher factor with less portfolio turnover leads to more lack in the trading strategy. Therefore, it is essential to examine multiple days' performances instead of considering only the current day or 1-day delay. If a factor is tradeable on the fifth day, then all predictions, which are 5 days old at that point, have reversed and are now actually predictive of negative returns. Figure 14.14 demonstrates the normal situation's alpha delay analysis, where Fig. 14.15 represents the COVID-19 situation results. The delay axis represents the number of individual delay days where an artificial delay is induced in the factor. In other words, what is the specific and the total Sharpe ratio that the strategy held effective one different delay days?

On the one hand, in the typical situation, even with the 14-day delay, the factor can still trade on 14 days ago and would still get much value. The second and third plots represent the total and specific return. In other words, these returns are what give rise to this plot using just the four first days as a delay. The final results keep improving overtime for the 4-day delay over time. On the other hand, in the COVID-19 situation, various factors' prediction power is underperformed and produces enormous losses.

14.5 Discussion

At the beginning of 2020, most financial specialists assumed the 11-year bull advertise to proceed, only to be shockingly clarified of that assumption by the spread of COVID-19. Consequently, the Dow fell from record highs to the bear-market

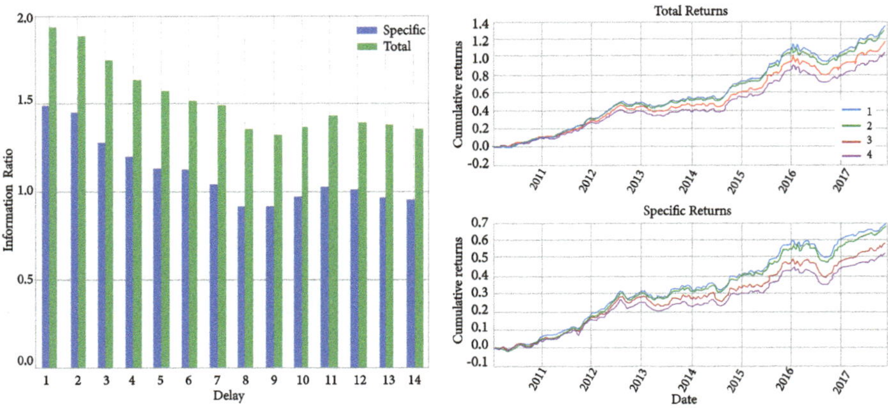

Fig. 14.14 2010–2018 situation analysis: Total Sharpe Ratio (IR) over the first 14 days in the alpha decay analysis (higher is better)

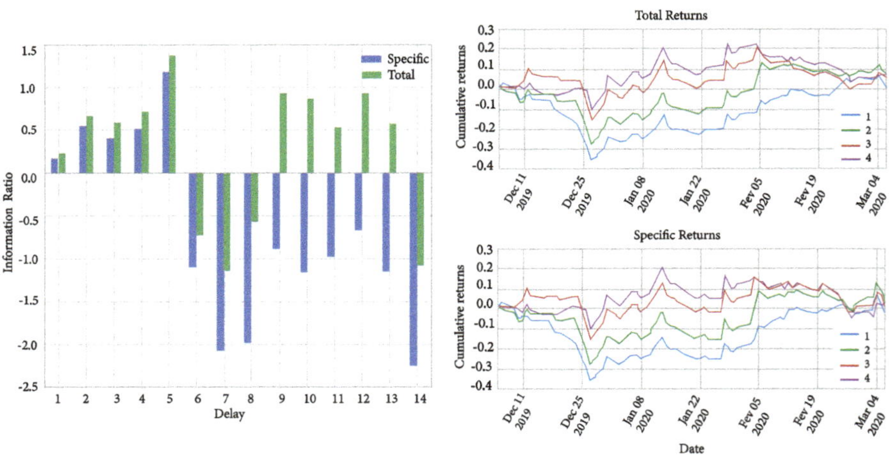

Fig. 14.15 COVID-19 situation analysis

region in weeks. Investors require a way to price in risk, and as of April 2020, numerous questions encompass COVID-19 for financial specialists to anticipate the economic effect, driving fear, and exceptional volatility. With market angst so high, it presses the investors to adopt policies that profit from volatility. The market has moved from high record to improvement region in a week and then exploded up 4.2% in a day. The impact of COVID-19 persists unknown. While continued volatility can be anticipated in a case such as this, it is wise to apply plans that improve returns, whether the market variations are violently down or up. Global trade will further get a punch, with worldwide fabricating sends out declining 50 billion in February alone. Stocks may be recovering from their March lows, but investor doubt over the pandemic's economic collapse is at extreme levels, with cash positions the highest.

Many quantitative companies have suffered significant losses during the subsequent selloff. However, three lessons could be learned amid the COVID-19 crisis. First, the developed system should provide more nowcasting and less forecasting. Forecasting relies on statistical relationships connecting lagged observations to future outcomes. Traditionally, quant strategies have focused on forecasting prices to gain long-range predictions. These strategies are based on price time-series dynamics and cross-sectional data (e.g., asset pricing and factor investing). Nowcasting models produce direct measurements, where the target variable is directly remarked (e.g., the basket of stocks utilized to measure inflation). Moreover, they allow short-range predictions, where the purpose variable is not directly observed (e.g., parking lot occupancy used to estimate revenue). The forecasting models allow the direct measurements that always hold (they do not rely on a statistical lead–lag relationship). Notice that, in both cases, the estimation phase involves millions of recent observations. Second, investors should develop theories instead of trading rules. In the current pandemic context, financial specialists are facing a decision-making problem under risk and pressure. Moreover, it is challenging to predict a probability distribution across the return and price. Therefore, the trader requires more than a price prediction approach or a rule-based policy that is not enough. Still, functional theories that explain a phenomenon by exposing a precise cause–effect mechanism are required. Finally, practitioners must avoid all-regime strategies. They search for investment strategies that would have performed well across all market regimes. Even if all-regime policies existed, they are likely to be a somewhat irrelevant subset of the population of strategies that work across one or more systems. Asset managers should focus their efforts on searching for investment strategies that perform optimally under specific market regimes. However, each regime is characterized by a particular data generating process (DGP). Hence, the investors can nowcast the probability, current observations being extracted from each DGP, and utilize them to build an ensemble portfolio of those optimal strategies. This chapter permits funds to adapt as market conditions change, e.g., from risk-on to risk-off strategies in the advent of COVID-19.

14.6 Conclusion

This chapter aims to inspect the financial influence of coronavirus. Hence, this chapter presents a starting point for many further interrogations. Different firm characteristics can be analyzed; credit and derivatives markets may have expected to come earlier. Banks may experience stress due to firms drawing on their credit lines, establishing which policies effectively improve that stress and differences across countries. Besides, it will further reveal the effectiveness and market perception of several policies. Despite the high loss in financial markets caused by the COVID-19 pandemic, the accumulated experiences before the stock-market chock can be extended carefully into the current situation. For example, momentum- and volatility-based strategy have been densely filled with insights to discover prominent relationships connecting the market performance and actions. Also, they improve regularly and continuously the quality of the results. Moreover, combining textual and financial data via sentiment analysis systems reveals how markets can respond to developments around COVID-19. The complete development of deep policy learning has to be the topic of coming investigations.

Appendix

Original and Customized Factors Utilized in this Work

Tables 14.3 and 14.4 highlight the original and customed factors, respectively. The Piotroski costumed factor consists of profit based on RoA and Cash Flow, shares outstanding operating based on gross margin and assets turnover, and leverages represented by long-term debt ratio current ratio.

Table 14.3 List of the factors

Type	Factor
Operation ratios	ROIC: Return on invested capital, current ratio
	RoA: Return on assets, EBIT margin
	Gross margin, assets turnover, quick ratio
Balance sheet	Total assets, Long-term debt
	Cash and cash equivalents, current debt
	Preferred stock, minority interest balance sheet
Valuation	Shares outstanding, enterprise value
Cash flow statement	Free cash flow, operating cash flow
Income statement	Net income continuous operations
	Cost of revenue, total revenue
Valuation ratio	FCF yield
Asset classification	Growth score, price/earnings to growth ratio (PEG)
Other	Altman's Z-score returns, sales - net - 1 Yr growth
	Capital Exp (Total) % Total assets
	Capital expenditures, volatility

Table 14.4 List of the Costumed Factors

Factor	Description
Sentiment Analysis	Twitter: Stocktwits
Direction	(close-open)/close
Mean revenue	Based on high, low, and close prices
Custom volatility	Based on high, low, close, and volume
Money flow Volume 5d	Based on close price and the corresponding volume
Trend line	Based on the closing price
Gross income margin	Gross profit divided by net sales
Max gap	The biggest absolute overnight gap in the previous 90 sessions
Capital expenditures	Capital Exp (Total) % Total assets
FCF EV	FCF × Shares outstanding/enterprise value
Debt to total assets	How much debt a company has on its balance sheet relative to total assets
TEM	Standard deviation of past 6 quarters' reports
Piotroski	Profit + leverage + operating
Advanced momentum	Momentum returns over 126
STA	$\frac{\lvert Net\ Income\ Continuous\ Operations - Operating\ Cash\ Flow \rvert}{Total\ Assets}$
ROA growth	1 if (roa > roa[252]) and =0 otherwise
FCF Total Assets (FCFTA)	1 if (FCF / Total Assets > 0) and =0 otherwise
FCFTA ROA	1 if (FCF / Total Assets > roa) and =0 otherwise
FCFTA growth	1 if (FCF / Total Assets > FCF / Total Assets (252)) and =0 otherwise
Long-Term Debt (ltd) GR	1 if (ltd/Total Assets <ltd[252]/Total Assets[252]) and =0 otherwise
Current ratio CR GR	1 if (cr > cr[252])and =0 otherwise
Gross margin (gm) GR	1 if (gm>gm[252])and =0 otherwise
Assets turnover (atr) GR	1 if (atr > atr[252]) and =0 otherwise
GM growth 2 YR	Geometric mean ([gm +1, gm[252]+1,gm[504]+1])-1)
GM stability 2 YR	Standard deviation([gm-gm[252],gm[252]-gm[504]],0)
ROA growth 2YR	Geometric mean([roa +1, roa[252]+1,roa[504]+1])-1)
ROIC growth 2YR	Geometric mean([roic +1, roic[252]+1,roic[504]+1])-1)
GM stability 8YR	Gross margin[8]
ROA growth 8YR	Geometric mean of [roa[i]/100 +1] / $i = 1 \ldots, 8$
ROIC growth 8YR	Geometric mean of [roic[i]/100 +1] / $i = 1 \ldots, 8$

References

1. A.M. Al-Awadhi, K. Al-Saifi, A. Al-Awadhi, S. Alhamadi, Death and contagious infectious diseases: impact of the covid-19 virus on stock market returns. J. Behav. Exp. Finance, 100326 (2020)
2. A. Atkeson, What will be the economic impact of covid-19 in the US? Rough estimates of disease scenarios. Tech. rep., in *National Bureau of Economic Research* (2020)

3. R. Baldwin, E. Tomiura, Thinking ahead about the trade impact of covid-19, in *Economics in the Time of COVID-19* (2020), p. 59
4. S. Barua, Understanding coronanomics: the economic implications of the coronavirus (covid-19) pandemic. SSRN Electron. J. (2020). https://doi.org/10/ggq92n
5. J. Bender, R. Briand, D. Melas, R.A. Subramanian, M. Subramanian, Deploying multi-factor index allocations in institutional portfolios, in *Risk-Based and Factor Investing* (Elsevier, Berlin, 2015), pp. 339–363
6. J. Bender, X. Sun, R. Thomas, V. Zdorovtsov, The promises and pitfalls of factor timing. J. Portfolio Manage. **44**(4), 79–92 (2018)
7. D. Blitz, M. Vidojevic, The characteristics of factor investing. J. Portfolio Manage. **45**(3), 69–86 (2019)
8. J.Y. Campbell, R.J. Shiller, The dividend-price ratio and expectations of future dividends and discount factors. Rev. Financ. Stud. **1**(3), 195–228 (1988)
9. G. Creamer, Y. Freund, Automated trading with boosting and expert weighting. Quant. Financ. **10**(4), 401–420 (2010)
10. M.S. Eichenbaum, S. Rebelo, M. Trabandt, The macroeconomics of epidemics. Tech. rep., in *National Bureau of Economic Research* (2020)
11. F. Eugene, K. French, The cross-section of expected stock returns. J. Financ. **47**(2), 427–465 (1992)
12. E.F. Fama, K.R. French, Common risk factors in the returns on stocks and bonds. J. Financ. Econ. **33**(1), 3–56 (1993)
13. E.F. Fama, K.R. French, A five-factor asset pricing model. J. Financ. Econ. **116**(1), 1–22 (2015)
14. L. Fornaro, M. Wolf, *Covid-19 Coronavirus and Macroeconomic Policy* (2020)
15. N.J. Gormsen, R.S. Koijen, Coronavirus: impact on stock prices and growth expectations, in *University of Chicago, Becker Friedman Institute for Economics Working Paper (2020–2022)* (2020)
16. R.C. Grinold, R.N. Kahn, *Active Portfolio Management* (2000)
17. S. Gu, B. Kelly, D. Xiu, Empirical asset pricing via machine learning. Tech. rep., in *National Bureau of Economic Research* (2018)
18. C.R. Harvey, Y. Liu, H. Zhu, ... and the cross-section of expected returns. Rev. Financ. Stud. **29**(1), 5–68 (2016)
19. B. Hirchoua, B. Ouhbi, B. Frikh, Rules based policy for stock trading: a new deep reinforcement learning method, in *Proceedings of the 2020 5th International Conference on Cloud Computing and Artificial Intelligence: Technologies and Applications (CloudTech)* (2020), pp. 1–6. https://doi.org/10.1109/CloudTech49835.2020.9365878
20. B. Hirchoua, B. Ouhbi, B. Frikh, Deep reinforcement learning based trading agents: risk curiosity driven learning for financial rules-based policy. Expert Syst. Appl. **170**, 114553 (2021). https://doi.org/10.1016/j.eswa.2020.114553. https://www.sciencedirect.com/science/article/pii/S0957417420311970
21. R. Huerta, F. Corbacho, C. Elkan, Nonlinear support vector machines can systematically identify stocks with high and low future returns. Algorithmic Finance **2**(1), 45–58 (2013)
22. A. Ilmanen, R. Israel, T.J. Moskowitz, A.K. Thapar, F. Wang, *Factor Premia and Factor Timing: A Century of Evidence*. Available at SSRN 3400998 (2019)
23. E. Jurczenko, *Factor Investing: From Traditional to Alternative Risk Premia* (Elsevier, Berlin, 2017)
24. D.P. Kingma, J. Ba, Adam: A method for stochastic optimization. arXiv preprint arXiv:1412.6980 (2014)
25. C. Krauss, X.A. Do, N. Huck, Deep neural networks, gradient-boosted trees, random forests: statistical arbitrage on the S&P 500. Eur. J. Oper. Res. **259**(2), 689–702 (2017)
26. B.G. Malkiel, E.F. Fama, Efficient capital markets: a review of theory and empirical work. J. Financ. **25**(2), 383–417 (1970)
27. W.J. McKibbin, R. Fernando, *The Global Macroeconomic Impacts of Covid-19: Seven Scenarios* (2020)

28. S. Ramelli, A. Wagner, What the stock market tells us about the consequences of covid-19, in *Mitigating the COVID Economic Crisis: Act Fast and Do Whatever* (2020), p. 63
29. S. Ramelli, A.F. Wagner, *Feverish Stock Price Reactions to Covid-19* (2020)
30. K.C. Rasekhschaffe, R.C. Jones, Machine learning for stock selection. Financ. Analysts J. **75**(3), 70–88 (2019)
31. C. Sohrabi, Z. Alsafi, N. O'Neill, M. Khan, A. Kerwan, A. Al-Jabir, C. Iosifidis, R. Agha, World Health Organization declares global emergency: A review of the 2019 novel coronavirus (covid-19). Int. J. Surg. **76**, 71–76 (2020)
32. J.H. Stock, Data gaps and the policy response to the novel coronavirus. Tech. rep., in *National Bureau of Economic Research* (2020)
33. J. Su, H. Fang, Research on multiple-factor quantitative stock selection strategy based on CSI 300 stocks. J. Fujian Bus. Univ. **1**, 21–28 (2018). https://www.doi.org/10.19473/j.cnki.1008-4940.2018.01.003
34. A. Subrahmanyam, The cross-section of expected stock returns: what have we learnt from the past 25 years of research? Eur. Financ. Manage. **16**(1), 27–42 (2010)
35. V. Vapnik, *The Nature of Statistical Learning Theory* (Springer, Berlin, 2013)
36. U. Von Luxburg, A tutorial on spectral clustering. Stat. Comput. **17**(4), 395–416 (2007)

Chapter 15
An Artificial Immune System for the Management of the Emergency Divisions

Mouna Berquedich, Ahmed Chebak, Oualid Kamach, Oussama Laayati, and Malek Masmoudi

15.1 Introduction

Numerous studies revolving around expert system have been conducted in medical diagnosis across the last decennary. Kellermann employed the Assistance System Clinical Decision (SADC) to reason in the vast medical fields and complexes [1]. SADC helps physicians to choose appropriate treatments and make the optimal decisions to enhance the performance of medical diagnoses. As Fig. 15.1 displays, the majority of SADC systems are constituted based on three segments: the premiere segment represented by knowledge base, which contains the rules; the second segment represents an inference engine, which employs logical rules of the knowledge base to figure out novel knowledge; and finally, the third segment that is a communication instrument showing the outcomes to the manipulator. Reasoning is a crucial function accomplished by the engine interpretation that is associating the medicative component of the SADC, which combines the medical knowledge of the SADC component with distinct data patients to produce appropriate decisions to be made. Reasoning approaches offer powerful apparatuses and methodologies to manipulate knowledge regarding the choice of a decision, which accordingly helps to resolve the issue (i.e., crises). The reasoning mechanism within a medical diagnosis service department is complex. It must take into account many clues involving the history of the patient, present symptoms, diagnosis outcomes, therapies offered,

M. Berquedich (✉) · A. Chebak · O. Kamach · O. Laayati
Polytechnic Mohammed VI University, Ben-Guerir, Morocco
e-mail: mouna.berquedich@um6p.ma; ahmed.chebak@um6p.ma; oulaid.kamach@um6p.ma; oussama.laayati@um6p.ma

M. Masmoudi
Jean Monnet University, Laboratory of industrial engineering, Roanne, France
e-mail: malek.masmoudi@univ-st-etienne.fr

M. Ouaissa et al. (eds.), *Computational Intelligence in Recent Communication Networks*, EAI/Springer Innovations in Communication and Computing,
https://doi.org/10.1007/978-3-030-77185-0_15

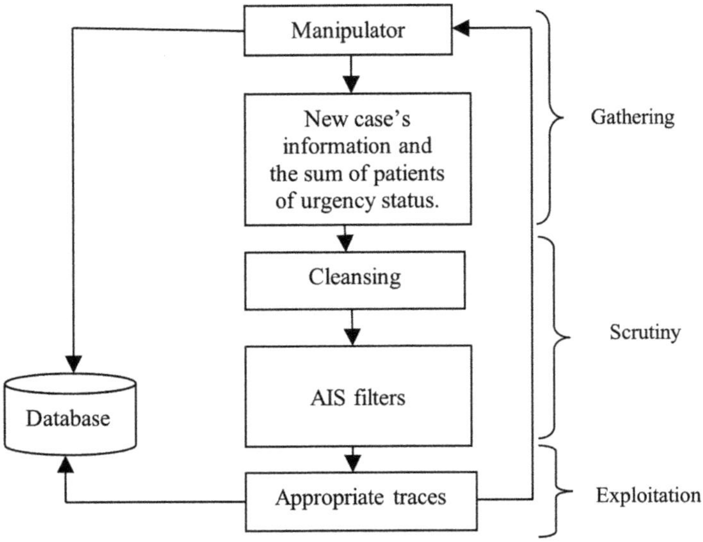

Fig. 15.1 Skeleton of the system

potential allergies, and so on. All these are provided to map them as well and assign them with a list of possibilities corresponding diagnostics. There exist various appropriate reasoning methodologies to variate displays of knowledge and fields of application [2, 3]. Actually, the case-based reasoning (CBR) and reasoning based on rules (RBR) are the most recognizable reasoning techniques [4].

15.2 Review of the Literature

15.2.1 Disruptions in Healthcare Facilities

Emergency divisions are critical facilities within hospitals [5]. They frequently face occurrences and/or unusual situations, that is, the augmenting numbers of patients as a result of epidemic incidents (i.e., explosions, massive accidents, natural catastrophes, and forth), shortage in resources, and the emergence of perplexing illnesses that demand considerable time disposition.

The results of such disturbances at the emergency divisions of hospitals may change from a natural peak of function to a condition of crisis and passing over conditions of tension and stress [6].

15.2.2 The Status of Stress Within the Emergency Divisions

Stress statuses may be unpacked based upon various viewpoints. It is hedged by multiple patterns. Based upon flow patient perspective, a status of stress in an emergency division may be seen as an overloaded pressure over the capacity of caregiving where the hedging rate is bypassed. In other words, a situation of pressure in healthcare emergencies is determined as unbalance in between care load and care capacity that is depicted through the excess of the upper marginal values. In one hand, the care load consists of the number of patients be it entering, outgoing, runaways, and the ones quitting the emergencies without being treated. On the other hand, the care capacity refers to the number of patients handled by emergency division in a given period of time, that is, taking into account material and human resources. This comprises the available number of nurses, doctors, caregivers, auxiliaries, beds, boxes, rooms, equipment, and medics, and it is anticipated by staff of emergency divisions.

Considering these patterns, and especially while dealing with the care capacity, this chapter defines two typologies of capacity. First, the required capacity, which is securing the resources functioning at maximal capability. Second, the capacity available, which is the results of a subtraction of the resources from the time of dysfunction.

The major patterns that may influence the balance are basically of two categories: firstly, patterns impacting the amount of influx (flows of patients) as the epidemics of seasons (i.e., influenza, bronchiolitis, flue, cold, gastroenteritis, and so on) and/or accidents (i.e., road, work, and manufacturing accidents) and secondly, patterns affecting the faster healthcare and hence the capability to generate cares that are either external or internal capability of transfer or both of them, that is, the disposition of subsequential healthcare services, etc. [7].

15.2.3 Improvement Actions to Deal with Situations of Tension

Hospitalizing victims of massive inflows is taken into consideration in contemporary ministerial publication renown as "Plan Blanc" (English: white plan). It presents the echelons of arrangement established in each private and public healthcare facilities (albeit along or beyond urgencies from these facilities). The plan aims at ensuring the optimum reception of patients taking into account the well-being of people that are under hospitalization. A cell of crisis (e.g., crisis unit) represents a key skeleton of the "Plan Blanc." As a result, it has been decided based upon formal arrangement planned in reflection sheets, that is, the creation of assembly points for patient to gathering victims and their relatives, media, families, and forth, and, too, to create pathways to these gathering locations and the possibilities to augmenting the capacity of addition into the healthcare entities as well as the control of the people suffering from illnesses [8].

15.2.4 White Plan: The Guiding Standards

A situation of healthcare crisis triggers huge number of synchronous victims. For long time gone by, it brought about policies of straightforward transfer of victims to the closest healthcare entities. This may be seen as "primitive" kind of transportation. By cumulation of multiple situations, these "primitive" used evacuations happened employing private ambulances or vehicles. Most of the time, they transfer the crisis from one area to another one. In fact, the unawarded and insufficiently prepared healthcare facilities were oftentimes the areas that Huguenard describes as "[. . .] an improvisation, a sterile agitation, multiple and contradictory orders, useless good will, unhealthy curiosity" [9]. This fully inappropriate and unsuitable receiving protocols may cause paralysis affecting the usual function within the healthcare facilities. The foundational standard of Emergency Plan is hence to offer an account to better receiving of the diseased patients, as long as ensuring the safety and security of the patients under ongoing care. It also facilitates and/or ensures various natural operation withing emergency divisions. In the field, how can these emergency divisions manage the increasing number of victims arriving at place?

This chapter emphasizes particularly the enhancement of the capacity of reception through the coordination in-between-ness of all divisions of healthcare facilities. It also focuses on augmenting the interchange of data through networks connecting different hospitals while preserving the safety of other patients under hospitalization that already exist within these hospitals.

15.2.5 Improving the Capacity of Reception Within Hospitals

The white plan explains the manner by which a hospital capacity may be increased with. This is possible by beginning with a census of the present means and equipment (i.e., rooms of operations, beds, laboratories, caregivers, and forth) as fast as it is affordable at appropriate time while the warning is declared. It follows that numerous processes kick off. That is, firstly, to preserve the personnel at site, specifically when an event happens while shifts interchange. Secondly, it follows that the decision-makers have to deprogram and lessen the functionalities of the operational areas, laboratories, and department of radiology. Thirdly, it is recommended to transfer the patients whose health conditions are stable or require lower caregiving and move them even to other healthcare units. Finally, it follows that the staff must be recalled by virtue of a tested system of multiplication; a graduation based upon the forecasted period; a record of up-to-date, trustful, and ready-to-use contacts' data; a refresh of their service or a particular epidermic spot; and an appropriate activity of the personnel and child nurseries.

15.3 Artificial Immune System Techniques: An Overview

15.3.1 Key Conception

The artificial immune system (AIS) domain was released based upon an inspiration from the natural immune system that numerous living species have. The driven idea was to invent systems that function in conditions identical to obstacles that the natural immune systems face. For De-Castro and Timmis, the AIS are "[...] the adaptive systems, inspired by the theories of the immunology, as well as the functions, the principles and the immune models, in order to be applied to the resolution of problems" [10]. In fact, the immunity is of a dichotomous subdivision, in other words two distinct systems that form the immunity: the adaptive and innate immune systems. The former type bears three organic methodologies, that is, clonal selection, negative selection, and immune networks, which form the adaptive immune processes' essence [11]. Noteworthy here to mention that the natural dendritic cells connect the adaptive immune system to the innate immune system.

15.3.2 Negative Selection

The reason behind using negative selection lurks beneath the tolerance it provides for self-cells [12], that is, the thymus, which is the way or gatekeeper versus non-self-antigens. The T cells that represents the non-self-antigens are smashed within this process (i.e., organ). The whole T cells withdrawing from thymuses and mobilizing in the corpus are considered to being tolerable of the self.

15.3.3 Clonal Selection

The clonal selection algorithm is embedded within the natural immune system to determine the elemental features of an immune responsiveness against antigenic stimuli [13]. This algorithm helps to make sure that solely the cells are able to recognize the antigens identified to fight against the intruders. These identified cells experience affinity procedures that enhance their discerning affinity toward antigens. They bear identifiers (i.e., detectors) within their own clusters. These identifiers are able to detect particular antigens at various levels of clonal selection (look at article [14] for further details).

15.3.4 Immunity Networks

The immunity networks conception has been primarily refined by Jerne [15]. It is a bioinspired computational model that utilizes the ideals and/or theorem of immunity network. That is mainly the interlay amid the cloning mechanism and B cells. It collects an antigen to be input and transmits it back as an inoculated network mixture of immunity B cells that are refined and addressed amid each other. The immunity network mechanism is relatively as similar as clonal selection in how it functions. However, there is a process of suppression that gets rid of cells presenting a kind of inception of affinity in between one another [16. 17].

15.3.5 Training Methods

Two kinds of cells are embedded to identify the present pathogen: B cells and T cells. Indeed, the population of these existing cells within the blood flow is crucial in identifying and fighting against the pathogenic organisms. The population reacts in collective manner and is able to recognize the novel pathogens, thanks to both clonal and negative selections [18]. During the phase of negative selection, the natural immune system (NIS) is capable to shield the tissues or the hosting organism from being invaded by its proper system of immunity. Part of the cells produce identifiers that identify proteins existing on cells' surfaces. The identifiers, renown as "antibodies," are generated randomly. Prior to being mature, these cells are "tested" within thymuses. These thymuses are organs situated beneath sternums and are capable to break up every immature cell that is depicted as tissues of non-self-organism [19]. The negative selection phase outlines the negative area of a likely class, that is, disposed instances of the class of "self." Historically, the algorithm of negative selection emerged back to 1994 [20]. By employing clonal selection, the NIS is able to address itself to afford the most valuable reaction face of the attacks by pathogens. This selection occurs as soon as any cell identifier detects a formerly recognized pathogen within the organism, which immediately clones itself to begin the immune reaction. Cloning systems, however, add little divergences in the order that the cell identifiers identify. The quantity of clones generated by a cell identifier vary articulating upon the affinity of the novel pathogen's cell. That is a way of measurement to identify how far the cell is able to mate the pathogen. The affinity, indeed, is negatively proportional to the quantity of deviation permitted, and the cells that present big chunk of affinities are mutated the least. Meanwhile, algorithms of clonal selection are as same as the natural selection mechanisms and hence are as same as genetical algorithms relying upon natural selectivity [21]. Notwithstanding, algorithms of clonal selectiveness have fewer parameters as compared to genetical algorithms, and possibly, they do not need complicated operation procedures. Most of the time, they are utilized as algorithms of optimization, and a little of them are employed in classification processes, and their first appearance went back to 1959 [22]. Further, the algorithms of artificial immune systems utilized for classifications are renown as classifiers as they gather the outcomes of various simplistic classifiers indeed.

15.4 Scrutiny of the Arriving Patients

The emergency division (ED) of hospitals is a service that requires organization given the nature of its activity. The flow of patients arriving at the emergency department varies from 1 day to another. This section will discuss the existing methods of organizing patients' flows at the emergency department scale. Various studies analyzed the control of patient inflows that randomly arrived to healthcare facilities articulating upon various techniques. For instance, Tandberg and Qualls employed triadic statistical approaches that are basically two seasonal decomposition methodologies and moving average method. They forecast the ED presentation value at every time of the weekdays within the university hospital of New Mexico [23]. Accordingly, Rotstein and other analysts developed statistical models for emergency divisions based in Israel [24].

Researchers as Abdel-Aal and Mangoud employed two univariate time series scrutinizing to forecast and model the number of patients on a monthly basis for over a decade (1986–1996). That was at the university of King Faisal, clinic of family and community medicine, and healthcare of Al-Khobar, Saudi Arabia [25]. Other researchers utilized Box-Jenkins models to anticipate entries to the emergency services on a daily basis and the rate of urgent cases that usually fill rooms and beds at Bromley healthcare facilities, NHS Trust, in the UK [25]. The work of Jones, Joy, and Pearson emphasized ED models in healthcare systems at Tenerife, Spain, and their study articulates upon ED's time series, specifically the hourly presence number span across 6 years (1997–2002). They fractioned the time series models into two categories: linear and nonlinear models [26].

Some work analyzed the utilization of knowledge-based data relying upon prior executed resolutions and prior scenarios to make appropriate decisions while managing massive influx of healthcare patients. The aim here is to provide a proactive method to resolve this issue.

15.5 Prioritizing Patients at Emergency Divisions

Triage method is made to sort the arriving patients based on their health conditions. This is important especially when it comes to the emergency divisions' echelons. In fact, five classification scales of severity had been put into action among the arriving patients to the ED. Every scale is assigned with measurements relative to the severity of every case. The available sorting tools that are available are as it follows.

Firstly, the emergency severity index (ESI), which is a five-point score hint that was developed, thanks to American research firm previously recognized as the Agency of Healthcare Research and Quality. This endorsed record carried various alterations that make it feasible to acquire the actual adaptation of 5 scores [27, 28]. Scoring 1 means that the patient is in unstable condition, which is of high degree of severity that requires an urgent intervention. In contrast, a patient scoring 5 means that their health condition is stable and does not need immediate intervention and/or emergent care.

Secondly, the Manchester Triage Scale (MTS) is another scale that was developed roughly toward the end of 1996. This scale proved its ability to detect people of severe health condition while arriving as patients to ED [29]. Object of better usage, involving training the medical staffs, the MTS has optimal sensitivity. This work depicts patient scales starting from MTS1 up until MTS5.

Finally, the Canadian Emergency Department (CRTC) is a computerizing triage echelon developed and was primarily incepted in 1998 by the Canadians before being largely employed in the United States of America [30, 31]. It takes into consideration the speed of the application of the caregiving and their time of reevaluation to implement the usage of care. It depends upon a normalized list of rationales to consulting along the conception of identification renowned as the first and second structure (i.e., order) CIMU echelons, which are distinctive and of gradual scales moving from 1 to 5.

It is for these reasons that an examination of the appropriate trace related to three scales had been selected, that is, scale 1 referring to urgent patients, scale 2 indicating moderate case of urgency, and scale 3 pointing at the nonemergent patients.

15.6 Projection of Filtering Methodology

Considering the obstacles, the issues stated above in the introducing section, and articulating the AIS methodologies employed, especially clonal and negative selections, this section describes the proactive process skeleton and the algorithm of filtration.

15.6.1 Overview of the System

The aim of this work is to develop integral supplying instrument for healthcare facilities that helps decision-makers to enhance the quality of the decisions they make before a situation of massive influx of patients. The crucial notion is to depict traces within databases and help board of directors and executives to recognize harmful scenarios employing the AIS techniques, mainly clonal and negative selectiveness. The correlation amid the standard of natural immune system and the

Table 15.1 Correlation amid system developed and natural immune system standard

Natural immune system (NIS)	Artificial immune system used in healthcare's emergency environment (AIS)
Clone and memorize.	Clone and refresh the data of self.
Identify the antigen.	Depict the inappropriate trace.
Lymphocyte (B cell and T cell)	Identifiers
Non-self (antigen)	Inappropriate trace
Self (antibody)	Appropriate trace

problematic as suggested in Table 15.1 prompted the research team to develop the system.

15.6.2 Distance Metrics Function

A distance or metric function is a function $d(x, y)$ that determines the similitude amid two distinctive vectors, that is, $x_i = [x_i1, x_i2, ..., x_id]$ and $x_j = [x_j1, x_j2, ..., x_jd]$. It grants a positive real number and displays how big the dissonance is amid two data points [32]. Despite the large amount of distance metrics suggested in the review of literature, the largest metric employed remains the Euclidian distance. This latter was refined by Euclid for over 2000-year gone by. Another widely utilized metric is Manhattan that is also renown as the distance of city block [32].

15.6.3 Choosing Manhattan Distance

Analysts as Noel and Bernardete presented a research that analyzes the influence that specific metric distance has in event-based algorithms of learning [33]. Particularly, they investigated the algorithm of nearest neighbor (NN) alongside the algorithm of incremental hypersphere classifier (IHC) that was efficient while implementing it in wide kinds of e-learnings and problem-solving recognition. They offered an account to empirically examine in details 15 datasets of various dimensions and shapes. Statistically speaking, they demonstrated that the Manhattan and Euclidian metrics generate relevant outcomes while being under cases of large-scale issues. Notwithstanding, methodologies of grid search are oftentimes demanded to identify the optimal matching metric relying upon the algorithm and the issue. They came out with conclusion that the distance of Manhattan is endorsed, especially while dealing with big datasets, as it required fewer computation as compared to other distances. That is what encouraged the research team to employ this methodology.

15.7 Manhattan Distance: The Matching Method

15.7.1 *Defining the Problem*

The supervised learning arrangement dilemma is identified in this part taking into consideration the data resolutions in the base of knowledge involving combinations of intakes and outtakes.

$$S = \{(t_1, y_1), (t_2, y_2), (t_3, y_3), \ldots, (t_s, y_n)\} \tag{15.1}$$

Given that t_i represents a vector of parameter rates at dimensional space *S*,

$$t_i \in T \tag{15.2}$$

A function which turns every vector t_i into space vector onto a class of *Y*-set, where.

$$T_i \in Y \tag{15.3}$$

The reason beneath implementing the algorithm of learning is to form a function *F* that approximates another function *E*, alongside *F* that is the true function that sorts an intake vector in *T* while under *Y*-type identified as.

$$F(t) : T \to Y \tag{15.4}$$

To examine the reliability of the model, sets of tests of data *T* have been utilized from an already present base of knowledge, that is,

$$Ti = \{t_1, t_2, t_3, \ldots, t_n\} \tag{15.5}$$

$$Tj = \{t_1, t_2, t_3, \ldots, t_n\} \tag{15.6}$$

$$W = \{w_1, w_2, w_3, \ldots, w_n\} \tag{15.7}$$

The vector *W* is the correlated weighing vector to *T* performance in the decision-making process. Otherwise, the distance of Manhattan is written as follows:

$$d\left(T_i, T_j\right) = \left|\sum \left(t_k\left(T_i\right) \text{-} t_k\left(T_j\right)\right)\right| \tag{15.8}$$

The mechanism of priorities is presented as follows:

$$\text{Min}\, d\left(T_i, T_j\right) = \min \frac{\Sigma\left(\left(tk(Ti) - tk\ (Tj)\right).wi\right.}{\Sigma wi} \tag{15.9}$$

15.7.2 Framework of Optimization

Within various healthcare facilities, it is beneficial to move nonurgent cases to another areas and ensure the presence of healthcare resources within these facilities rather than coexisting with the insufficient needs and/or long waiting periods inside of it. Consequently, by depicting the optimal amount of resources needed, considering the transfer of patients' proposition, this will decrease crowds within healthcare environments. The model of optimization contemplates the elemental capacity of healthcare facilities. Here are the mathematical sets of parameters used in this model:

c_i: Cost of raising a capacity unit at hospital i
d_i: Overtime cost within hospital i
$d_{i,l,t}$: Sum of type 1 patients discharged from hospital i in time t
H: Total number of healthcare entities
i, j: Indicators of healthcare entities and areas of need
$k_{i,t}$: Accessible volume in hospital i in time t
l: Patients' urgency indices (1 = urgent case, 2 = moderate urgent, 3 = not urgent)
L: Divergent-type patients
$n_{i,j}$: Count of nonurgent patients (l = 3) transferred from hospital i to hospital j
$p_{i,k,t}$: Count of type 1 patients received in hospital i in time t
t: Indicators of time
$\underline{x}_i$: Capacity regarded at hospital i

$$\text{Min} \sum\nolimits_{i \in H} c_i x_i + \sum\nolimits_{i \in H} \sum\nolimits_{j \in H} z_i, jn_i, j \tag{15.10}$$

Considering the constraints of capacity, the all-time accessible volume is identical to the preceding capacity that varied in accordance with the patients being either permitted or discharged.

$$k_{i,t} = ki, t\text{-}1\text{-}\sum\nolimits_{l \in L} p_i, l, t\text{-}1 + \sum\nolimits_{l \in L} d_i, l, t \tag{15.11}$$

The sum of the received patients ought not to bypass the capacity in hands.

$$\sum p_i, k, t \leq ki, t \forall i \in H \tag{15.12}$$

Capacity at disposition has to be less than or equal to the resources assigned to the specific and concerned shift team.

$$ki, t \leq xi\forall i \in H \tag{15.13}$$

The restrain to transferring patients has to be lesser than the limit of transfer of the initial healthcare facility and the capacity at hands in the targeted hospital.

$$ni, j, t \leq kj, t, j\forall i, j \in H \tag{15.14}$$

The skeleton of the system at work and the identification of the algorithms employed are presented in the following section.

15.8 System Skeleton

As pointed out by several studies, the treatment of traces experiences three major stages, that is, the gathering, scrutiny, and exploitation. This work emphasizes the second and third stages particularly. Under the scrutiny (or analysis) stage, the traces gathered in the first stage are stored and filtered, deploying the distance of Manhattan [33]. It follows that in the last stage, which is the exploitation phase, the aim is to afford the board of decision-makers with relevant traces and to maintain all suitable learning traces within the base of knowledge.

15.8.1 Gathering Traces

The present system combines the whole activities made by the board of decision-makers collectively. The traces recorded in the base of knowledge are retrieved from the manipulator-program interplay. Noteworthy here to mention that other traces can happen within other servers the system granted the access to.

15.8.2 Scrutinizing Traces

Considering the form of the trace, a novel dataset is generated as long as traces are in reformulation in accordance with the suggested format. The global trace's format is described as follows (Fig. 15.2).

$$\mathrm{T} = \{t_1, t_2, t_3, t_4, \ldots\}$$

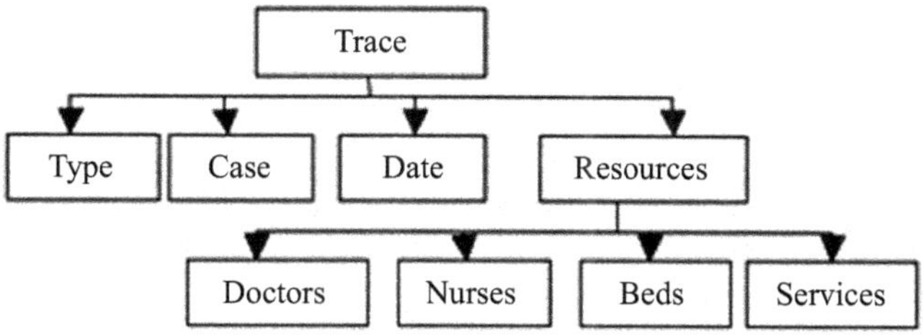

Fig. 15.2 Trace format

The letter *t* refers to the typologies of cases, which is featured by four tuples.

$$t = \{T, C, D, R\}$$

where.
C: Case representing the count of patients and their illness
D: Date
R: Resources, which is calculated as

$$R = \{Doc, N, B, S\}$$

T: Typology

15.8.2.1 Basic Concept of Negative Selectivity NSA

The cleansing phase aims at eliminating all sort of noise, that is, the sum of patients that their health conditions do not correspond to the case input. This possible, thanks to the cleaning algorithm embedded which cleans up all the traces and presents as follows (Fig. 15.3).

Algorithm 1: Cleansing traces output: CT = a set of cleansed traces

```
ND: Nearest distance
NT: set of nearest T
begin
For all (trace t ∈ T) then
Ct: case of trace.
D: d(Ct, C) = |C-Ct|
if (ND is null or D <ND) then
ND = D
t ∈ NT
else if (d = ND)
t ∈ NT
end for
```

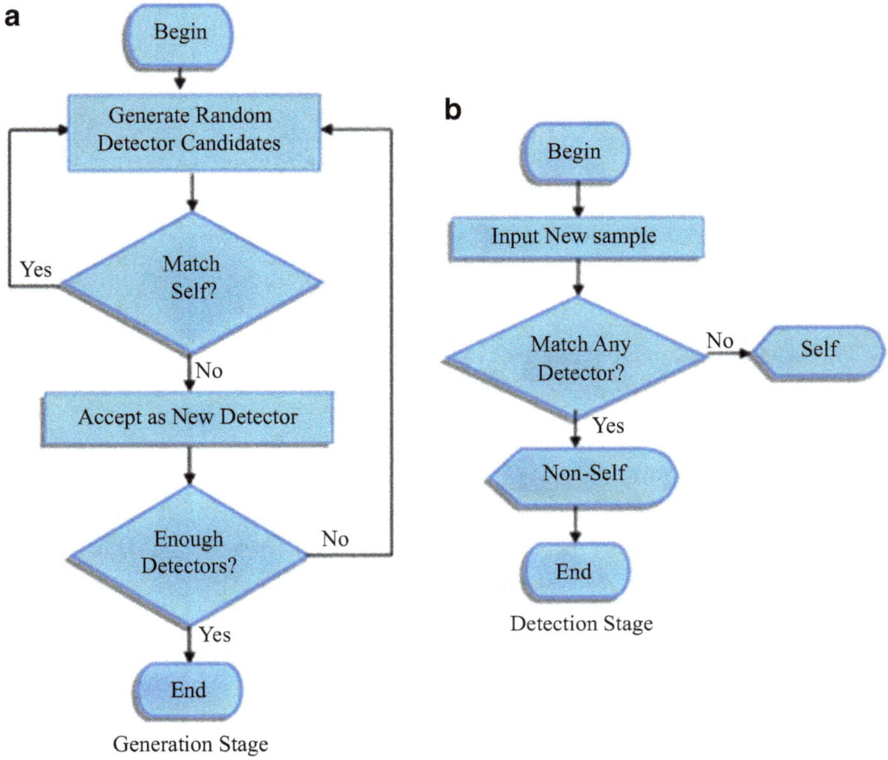

Fig. 15.3 Generation and detection stage

end

The filtering phase shall aim to automatically identify scenarios of disturbances that already existed and/or recorded versus the real arrived case. This is realized by the decision-makers. This work utilizes the algorithm of filtration to make triage of multiple traces and conserve solely the appropriate elements that will engender pertinent traces to the case provided by the decision-makers.

15.9 Outcomes of Computation

Looking forward guaranteeing the validity of the work's ideals, this project sets up a program named Emergency Management (EM). The main objectives of this system lurk beneath filtering the entire traces existing in the database and displaying solely the traces of pertinence of the cases that had been entered at the interface of the system (Fig. 15.4).

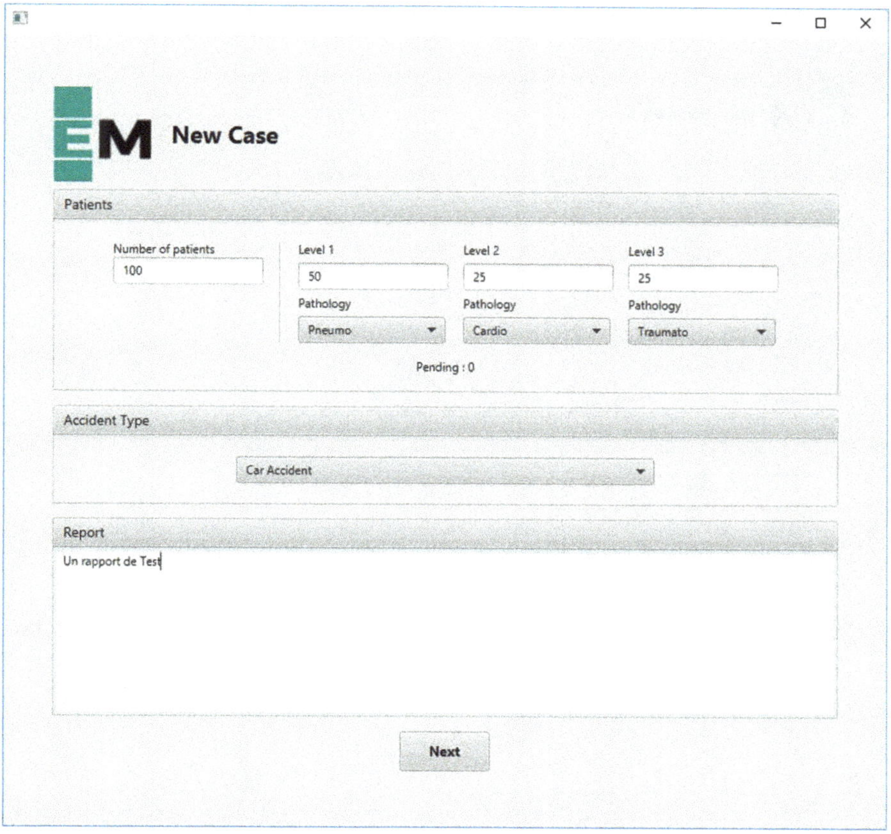

Fig. 15.4 Outcomes of the suggested algorithm to identify pertinent resolutions

The forecasted program looks at the matching incident solution in the database taking into account the resources at hand, and it coordinates through connectivity networks with different databases analogous to territorial healthcare facilities. Moreover, the phase of selection of the resources considers the means that exist in the healthcare facility that represents the host in real time. Simultaneously, it cooperates with other healthcare facilities in the region. The priority of selectiveness of the resources in every hospital is presented, counting the cost and constraints of the budget that emerge while transporting patients, constraints of logistics, and the obstacles of the capacity at disposition in every health facility. The figure below displays the outcomes collected from the program (Fig. 15.5).

The outcomes of dispersion related to the patients are presented based on their urgency records, that is, level of emergency, which hold constraints respective of capacity, budget, and logistics (Fig. 15.6).

This project employs AIS methodologies in healthcare environment. Its aim is to supply the board of decision-makers with pertinent traces, which looks at identical

Fig. 15.5 Displaying optimal solutions

circumstances in historical scenarios and/or records of cases already experienced, to offer them with the pertinent components. The reason of this is to enhance the scenarios at real time and afford the directors with minimal but significant traces and datasets. The process of filtration incorporates AIS methodologies in that clonal and negative selectiveness are integrated, that is, the clonal selectiveness that offers an account to make the original database up to date and the negative selectiveness that is deployed to identify the pertinent traces within it. The subsystem developed identifies the entire appropriate arrays of functions and advocates the board of decision-makers to depict and decide effectively.

This project, too, articulates the standards of white plan that was subjected to computation to bring real-time reactions and make it accessible to decision-makers to distribute the arriving patients based on their degrees of emergency. The cooperation among healthcare entities admit constraints related to volume, budget, and logistics, as well as preserving the safety and security of people under hospitalization in these entities. To meet the white plan's specification limits, the developed system offers an "all-risk" strategy in emergency divisions.

Fig. 15.6 The patients' dispersion across territorial hospitals counting their urgency levels

15.9.1 Alimenting the Database

The alimentation of the database was met by scenarios that were created randomly. That is, 20 cases were chosen where for each number of patient(s), there existed 50 traces representing diverse degrees of gravity and suffering from diverse illnesses. To optimize the results of selection, this project uses algorithms of filtration and memorization.

To ensure the viability and the accuracy of the employed approach, the final application uses a database of 10,000 incidents. The outcomes of the execution are displayed in the table below (Table 15.2).

After running the system, a sample size was selected to validate the implemented approach. Then the time of execution in each case was given. Under the first instance, the count of patients was 90, and the targeted resolution was run in the internal database. That was, too, across the same counts with diverse levels of gravity and illnesses. Then in the second instance, the count of patients was 60,

Table 15.2 Presentation of sample retrieved from database

Hospital	Number of patients by case	Count of traces by sum of patients	Count of degrees of severity by trace	Utilized resources number by group of severity
1	60 70 80 90 100 110 120 130 140 150 160 170 180 190 200 210 220 230 240 250	Beyond 50 traces by count of patients showing various sequences	3	Minimal one local resource. Maximal three members where one is of internal resources in addition to two resources from extern.
Sum of hospitals	**Total of cases 20**	**Sum of traces that exceed 1500**	**Sum of groups exceeding 3000**	**With 9000**

while the targeted resolution trial was due in two separate databases that represented different identical scenarios of similar count of patients. However, this time, they were of various types of pathologies and gravities. And finally, the third instance was run to depict resolution from one intrinsic database and two extrinsic databases that represents two external healthcare entities selected in this instance. Similarly, that is featured by the existence of pairs of identical scenarios while of differences in levels of gravity and illnesses.

Noteworthy here to mention that the load time refers to the required time to load the data from a hospital and hospitals cooperating in the dispersion of resources. This was implemented one single time during the run of the application, which functioned and eliminated the repeated needs of data to be retrieved frequently from the main databases and the subsequent ones.

15.10 Conclusion

The field of artificial immune systems (AIS) is relatively novel to bioinspired computing. The natural immune networks, and especially the biological models of it, together with negative and clonal selections, and immunity networks were behind the inspiration of AIS algorithms. These algorithms were deployed effectively across vast numbers of application and digital systems. The main argument of this project is to present a tool that is able to trace pertinent records to help the board of decision-makers to enhance their decisions and cope with challenges while executing their duties. This project also helps them to overcome the constraints resulted from their old solutions and explore appropriate data to meet their demands.

In the forthcoming days, it will be an interesting notion to employ the electronic health traces to identify healthcare entities' pathways. This points out at the sort of intervention that will change, specifically in the application, records, and treatment of data of every patient. In terms of approaches, this promotes the research team to suggest a methodology to develop the function within the healthcare facilities based on various techniques. In other words, the aim is to propose to retrieve general hospitals' pathways articulating upon data of health records, patient-based, and adaptable with the changing flow of patients within emergency divisions and that for enhancing the quality of care provided for patients in needs.

References

1. A. Kellermann, Crisis in the emergency department. N. Engl. J. Med. **355**(13), 1300–1303 (2006)
2. F. Harrou, F. Kadri, S. Chaabane, C. Tahon, Y. Sun, Improved principal component analysis for anomaly detection: application to an emergency department. Comput. Ind. Eng. **88**, 63–77 (2015)
3. F. Kadri, C. Pach, S. Chaabane, T. Berger, D. Trentesaux, C. Tahon, Y. Sallez, Modelling and management of strain situations in hospital systems using an orca approach, in *Proceedings of 2013 International Conference on Industrial Engineering and Systems Management, IESM*, (2013), pp. 1–9. IEEE
4. F. Kadri, F. Harrou, S. Chaabane, C. Tahon, Time series modelling and forecasting of emergency department overcrowding. J. Med. Syst. **38**(9), 1–20 (2014)
5. Early Detection of Abnormal Patient Arrivals at Hospital Emergency Department, in *6th IESM Conference*, (Seville, Spain 2015)
6. F. Kadri, S. Chaabane, C. Tahon, A simulation-based decision support system to prevent and predict strain situations in emergency department systems. Journal **42**, 32–52 (2014)
7. P. Carey, G. Cuthbert, R. Dang, B. Greystoke, A. McGregor, R. Oakes, J. Wallis, A more devolved and inclusive approach to integrated reporting facilitated by an IT system (Haemosys) networked to local information management systems (LIMS) in all participating regional hospitals. Br. J. Haematol. **173**(1), 43–43 (2016)
8. D. Carvalho, V. Joao, A. Rocha, J. Vasconcelos, Towards an encompassing maturity model for the Management of Hospital Information Systems. J. Med. Syst. **39**(9) (2015)
9. C. Virenque, Large influx of injured people in hospital. Hôpital Purpan, TSA 40031, 31059 Toulouse cedex 09, Journal, (2016), pp. 712–715
10. L.N. De Castro, J. Timmis, Artificial immune systems: a novel paradigm to pattern recognition, in *Artificial Neural Networks in Pattern Recognition*, ed. by J. M. Corchado, L. Alonso, C. Fyfe, (Springer Verlag, Paisley University, 2002), pp. 67–84
11. J. Timmis, A. Hone, T. Stibor, E. Clark, Theoretical advances in artificial immune systems. Theor. Comput. Sci. Rev. **403**(1), 11–32 (2008)

12. U. Aickelin, D. Dasgupta, Artificial immune systems, in *Research Methodologies*, ed. by E. K. Burke, G. Kendall, (Springer, New York, 2005), pp. 375–399
13. De Castro, L.N., Von Zuben, F.J. The clonal selection algorithm with engineering applications, in *Paper presented at The Workshop on Artificial Immune Systems and Their Applications*, (Las Vegas, 2000)
14. L.N. De Castro, F.J. Von Zuben, Learning and optimization using the clonal selection principle. IEEE Trans. Evol. Comput. **6**(3), 239–251 (2002)
15. N.K. Jerne, Towards a network theory of the immune system. Ann. Immunol. **125**, 373–389 (1974)
16. J. Greensmith, U. Aickelin, G. Tedesco, Information fusion for anomaly detection with the dendritic cell algorithm. Inf. Fusion J. **11**(1), 21–34 (2010)
17. J. Greensmith, U. Aickelin, S. Cayzer, Introducing dendritic cells as a novel immune-inspired algorithm for anomaly detection, in *Proceedings of the 4thInternational Conference on Artificial Immune Systems*, ed. by C. Jacob, M. L. Pilat, P. J. Bentley, J. I. Timmis, (Springer, Heidelberg, 2005), pp. 153–167
18. M. Burnet, The clonal selection theory of acquired immunity, in *The Clonal Selection Theory of Acquired Immunity* (1959)
19. A. Ala, G. Ajay, K. Dionysios, Optimizing an artificial immune system algorithm in support of flow-Based internet traffic classification Brian Schmidt (2017)
20. S. Forrest, A. Perelson, L. Allen, R. Cherukuri, Self-nonself discrimination in a computer, in *Proceedings of the 1994 IEEE Computer Society Symposium on Research In Security And Privacy*, IEEE (1994), pp. 202–212
21. L.N. De Castro, F.J. Von Zuben, Learning and optimization using the clonal selection principle. IEEE Trans **6**(3), 239–251 (2000)
22. M. Burnet, The clonal selection theory of acquired immunity, in *The Glonal Selection Theory of Acquired Immunity* (1959)
23. D. Tandberg, C. Qualls, Time series forecasts of emergency department patient volume, length of stay, and acuity. Ann. Emerg. Med. **23**(2), 299–306 (1994)
24. Z. Rotstein, R. Wilf-Miron, B. Lavi, A. Shahar, U. Gabbay, S. Noy, The dynamics of patient visits to a public hospital: a statistical model. Am. J. Emerg. Med. **15**(6), 596–599 (1997)
25. R. Abdel-Aal, A. Mangoud, Modeling and forecasting monthly patient volume at a primary health care clinic using univariate time- series analysis. Comput. Methods Prog. Biomed. **56**(3), 235–247 (1998)
26. S. Jones, M. Joy, J. Pearson, Forecasting demand of emergency care. Health Care Manag. Sci. **5**(4), 297–305 (2002)
27. D. Eitel, D. Travers, A. Rosenau, N. Gilboy, R. Wuerz, The emergency severity index triage algorithm version 2 is reliable and valid. Acad. Emerg. Med. **10**(10), 1070–1080 (2003)
28. P. Tanabe, R. Gimbel, P. Yarnold, J. Adams, The emergency severity index (version 3) 5-level triage system scores predict ED resource consumption. J. Emerg. Nurs. **30**(1), 22–29 (2004)
29. M. Cooke, S. Jinks, Does the Manchester triage system detect the critically ill? J Accid Emerg Med **16**(3), 179–181 (1999)
30. J. Jimenez, M. Murray, R. Beveridge, J. Pons, E. Cortes, J. Garrigos, et al., Implementation of the Canadian emergency department triage and acuity scale (CTAS) in the Principality of Andorra: can triage parameters serve as emergency department quality indicators? CJEM **5**(5), 315–322 (2007)
31. M. Bullard, B. Unger, J. Spence, E. Grafstein, Group CNW. Revisions to the Canadian emergency department triage and acuity scale (CTAS) adult guidelines. CJEM **10**(2), 136–151 (2008)
32. R.O. Duda, P.E. Hart, D.G. Stork, *Pattern Classification* (Wiley, New York, 2001)
33. L. Noel, R. Bernardete, *On the Impact of Distance Metrics in Instance-Based Learning Algorithms* (UDI, Polytechnic of Guarda, Guarda, 2017)

Index

A
Access control, 9, 10, 89, 91, 104
AdaBoost, 106
Advanced driver Assistance Systems (ADASs), 120
Alakrot's YouTube comments collection, 154–155
Alamouti decoding, 22
Anomalies/intrusions, 104
Anti-HIV drugs, 191
Antiretroviral (ARV) drugs, 191
Arabic social media, ML for hate speech detection in, 147–148
  classification performance scores, 159
  dataset, 160
    Alakrot's YouTube comments collection, 154–155
    YouTube, 153
  definition of, 148
  democratization of access, 149
  methodology
    evaluation metrics, 158–159
    Logistic Regression, 156–157
    Long Short-Term Memory, 158
    preprocessing, 155–156
    Random Forests, 157
    Support Vector Machines, 157
  MSA, 150
  natural language processing, 150–151
    Bag of Words and term frequency, 152
    stemming, 151
    term frequency-inverse document frequency, 152
  techniques, 149
Arabizi, 149
Artificial immune system (AIS)
  alimentation of database, 261–262
  clonal selection algorithm, 249
  definition, 249
  electronic health traces, 263
  Emergency Management (EM) outcomes
    clonal selectiveness, 260
    optimal resolutions, 260
    patients' dispersion across territorial hospitals, 261
    pertinent resolutions, 259
  filtering methodology
    distance metrics function, 253
    Manhattan distance, 253, 254
    system overview, 252–253
  gathering traces, 256
  healthcare's emergency, 253
  immunity networks, 250
  negative selection, 249
  scrutinizing traces
    cleaning algorithm, 257
    generation and detection stage, 258
    trace format, 257
  training methods, 250–251
Artificial intelligence (AI), 2, 10, 135
Artificial neural network (ANN), 106, 200–201
Assistance System Clinical Decision (SADC), 245, 246
AtomicProcess, 179
Attack-defense tree
  refinement and countermeasures, 71
  ROI and ROA, 74–76
  types of optimization, 71

M. Ouaissa et al. (eds.), *Computational Intelligence in Recent Communication Networks*, EAI/Springer Innovations in Communication and Computing,
https://doi.org/10.1007/978-3-030-77185-0

Attack-defense tree (*cont.*)
  VANET availability, 72–74
    black hole attack, 73
    denial of service attack, 73
    DOS attack, 73
    game theory, 76
    intrusion detection system, 73
    Nash equilibrium, 80
    players, 77
    reduction of, 79
    smurf attack, 73
    strategies, 77
    SYN cookies, 73
    transmitting dummy messages, 73
    utility matrix, 77, 78
Attribute-based signature (ABS), 84, 92–93
Authentication, 9, 10, 84, 89
Automatic composition, 167
Autonomous Pedestrian Crossing (APC), 135, 136, 145
  APC protocol, 138–139
Autonomous Traffic Management (ATM), 136

B
Bag of Words (BoW)
  natural language processing, 152
BioPortal dataset, 207, 210, 211
Biotech research, 231
Bit error rate (BER), 35–36
Blind source separation (BSS), 22, 23, 26, 27, 29, 33, 37
Blockchain
  applications, 2
  distributed ledger, 3
  features, 2
  oracle, 5
  sharding, 5
  smart contracts, 5
  taxonomy, 3
  technology, 2
  transactions, 3
  workflow, 3, 4
Blockchain consensus, 3, 4, 8, 11, 16–18
Blockchain integration with 5G networks, 19
  challenges
    infrastructure costs, 18
    interoperability, 17–18
    resource constraints, 17
    scalability and performance, 16
    security, 18
    standardization and regulations, 17
  conceptual diagram, 8, 9
  opportunities, 7–9
  performance enhancements, 10–11
  scalable and secure blockchain architecture, 11
  security improvements, 9–10
  shared services and operations, 8
  stakeholders, 8
  technologies, 7
Blockchain oracles, 5, 11, 13
Blockchain technology, 2
  characteristics, 3
  digital signature, 3
  hash function, 3
  shared services and operations, 8
  timestamp, 3
Block5G, 11, 12, 15
Box-Jenkins models, 251

C
Canadian Emergency Department (CRTC), 252
Case-based reasoning (CBR), 246
Cash flow, 240
Cell of crisis, 247
Cellular service providers (CSPs), 2, 137
Center and Scale Prediction (CSP), 137
Centralized cellular networks, 1
Centralized composition, 167
Channel State Information (CSI), 40, 50, 51
ChEMBL, 207
ChemSpider, 207
CISCA architecture, 183
Clinical patient data, 208
Clonal selection algorithm, 249
Cloud computing (CC), 1, 7, 9
Clustering algorithms, 194
Clustering process, 206, 209–210, 212, 213, 215
CompositeProcess, 179
Composition layer, CISCA architecture, 185
Computational assumption, 92
Confidentiality, 90
Confusion matrix, 111
Connected and Autonomous Vehicle (CAV) system, 119–120
Consortium blockchain, 4
Context Sensitive Systems, 164
Convolutional neural networks (CNNs), 49, 52, 53, 136
  convolutional layers, 56
  correction layers (ReLU), 57–58
  feed-forward neural network, 55
  fully connected layers, 57
  learning phase, 55

loss layer, 58
multilayer neural networks, 55
multiple filters, 57
vs. recurrent neural network (RNN), 61–62
shared weights, 56
structure of, 56
subsampling layers, 57
training process, 55
Cooperative awareness Message (CAM), 120
COVID-19 impact on stock markets
all-regime policies, 239
on companies according to geographic revenue, 235–237
containment policies, 220
cross-sectional investment, 223–225
cross-sectional investment-based clustering, 225–227
and customized factors, 241
delay, 237
COVID-19 situation results, 237, 238
normal situation's alpha delay analysis, 237, 238
forecasting models, 239
future economic impact, 219
growth analysis
healthcare growth analysis, 231, 233
insurance growth analysis, 233, 234
relative winning and losing US industries, 232
relative winning and losing US sectors, 232
incubation period, 220
investment factors, 222–223
on markets and economy
correlation degree between reported cases in different regions, 230
diverse clusters evolution to number of confirmed cases, 230
diverse critical situations, 228
SPY and SHY z-score-normalized minute dollar volume difference, 231
velocity of pandemic over different continents, 229
market returns, 220
original factors, 240
spectral clustering algorithm, 221, 225
stock return, 220, 221, 233
2008 crisis comparison, 233–235
visualization for mapping related areas, 236
Cross-sectional investment, 223–225
Cyberbullying, 147–148

D
DARPA Intrusion Detection Evaluation Program, 109
Data generating process (DGP), 239
Data management layer, CISCA architecture, 184–185
Data mining, 150, 193, 206, 207
Deep convolutional neural networks (CNNs), 106
Deep-learning (DL) techniques, 40
CNNs, 49
DL-based NOMA, 50, 51
DNN model, 48
for FBMC/OQAM, 22
for MIMO-FBMC/OQAM system, 29 (*see also* MIMO-FBMC/OQAM system)
NOMA method level, 40
RNNs, 48
Deep machine learning
conventional neural network, 139–140
long short-term memory, 140–141
Deep neural network (DNN), 48, 135
in MIMO-FBMC/OQAM systems, 29–33
RMSE and loss, 34–35
training, 30
Deep Q-learning, 120–121, 129
Deep Reinforcement Learning (DRL), 120
Device to device (D2D), 1, 7, 9, 11
Distant bicycle detection and avoidance (DBDA), 120
deep neural network architecture, 128–129
deep reinforcement learning, 122–123
Markov decision process modeling, 123–126
optimal driving policy estimation, 126–128
protocol, 121–122
Distributed composition, 167
Distributed Denial of Service (DDoS), 110
Distributed ledger technologies (DLT), 2
Doc2ec, 213
Doc2vec tool, 208, 209, 213, 215
Driver assistance functions, 53
Drug discovery, QSAR anti-HIV feature selection and prediction, 191–192
approach and validation
data, 195–196
proposed approach, 196–198
validation, 199–202
genetic algorithms, 193–194
machine learning methods, 194
Dynamic composition, 167

E

Electronic Health Records
  BioPortal dataset, 207, 210, 211
    F-measure, 214
    purity, 214
  covid disease dataset, 211
    F-measure, 214
    purity, 214
  diseases identification on clustering RDF data set, 209–210
  future work, 214–215
  linked data, 208–209
  validation metrics, 212–213
Electronic Medical Records (EMRs), 206
Elliptic Curve Digital Signature Algorithm, 3
Emergency division (ED)
  AIS (*see* Artificial immune system (AIS))
  disruptions in healthcare facilities, 246
  increasing hospital capacity, 248
  prioritizing patients, 251–252
  scrutiny of arriving patients, 251
  status of stress within, 247
  white plan, 248, 260
Emergency severity index (ESI), 252
Energy sector, 231
Enhanced mobile broadband (eMBB), 21, 39
Ethereum blockchain, 5
Euclidian distance, 253
Euler's constant, 109
European Debt Crisis 2011, 227
Experimental Factor Ontology (EFO), 207

F

Facebook, 147
Fama–French three-factor model, 221
Faster Region-based CNN (FR-CNN), 136
FBMC-OQAM (offset quadrature amplitude modulation-based filter bank multicarrier), 21
  multicarrier modulation technique, 23
  spectrum efficiency, 21
  symbol stream, 23
Fever period, 220
5G Automotive Association (5GAA), 88
Fifth-generation (5G) networks
  blockchain integration, 2
  challenging aspects, 2
  cloud computing (CC), 1, 7, 9
  5G new radio (NR), 21
  high-level performance targets, 39
  MEC, 6
  network slicing, 6
  NFV, 6
  road map, 1
  SDN, 5–6
  security challenges, 1–2
  technologies, 1
Financial markets, 222, 233, 239
F-measure, 212
4G, 1, 39–41
Fully Convolution Network (FCN), 137

G

GA-machine learning approach (GA-MLA), 196, 203
Game theory, 76
Genetic algorithm (GA), 192
  drug discovery, QSAR anti-HIV feature, 193–194
German Recognition Benchmark database, 63
German Traffic Sign Recognition Benchmark, 63
Global trade, 238

H

HAST-IDS, 106
Hate speech, 147–148
Hate speech detection, Arabic social media, 147–148
  classification performance scores, 159
  dataset, 160
    Alakrot's YouTube comments collection, 154–155
    YouTube, 153
  definition, 148
  democratization of access, 149
  methodology
    evaluation metrics, 158–159
    Logistic Regression, 156–157
    Long Short-Term Memory, 158
    preprocessing, 155–156
    Random Forests, 157
    Support Vector Machines, 157
  MSA, 150
  natural language processing, 150–151
    Bag of Words and term frequency, 152
    stemming, 151
    term frequency-inverse document frequency, 152
  techniques, 149
Head-Body Alignment Network (HBAN) model, 137
Healthcare crisis, 248
Healthcare firms, 231

High-level protocol specification language (HLPSL), 95
Human immunodeficiency virus (HIV) infection, 191
Hybrid architecture, 86–87
1-[(2-Hydroxyethoxy) methyl]-6-(phenylthio)thymine (HEPT), 192

I
IACA OWL-S (Intention And Context Adaptation OWL-S), 180
Immunity networks, 250
Incremental hypersphere classifier (IHC), 253
INRIX transport data company, 135
Integrity, 89–90
Intelligent delay, 222, 227
Intelligent transportation system (ITS), 83, 119, 135
Intentional-contextual service, 187
Intention and context modelling, 175
  context ontology, 177–178
  intention ontology, 177
  ontology-based intention and context modeller, 176–177
  proposed meta-model, 175–176
  semantic integration, OWL-S extension for, 178–181
Intention ontology, 177
Interference, 22
  imaginary interference, 24
  management, 10–11
  mitigation in MIMO-FBMC/OQAM system, 29–33
  NR interface, 40
  successive interference, 40, 42
11th International Classification of Diseases (ICD11), 207
International Mobile Telecommunications (IMT), 39
The International Telecommunication Union (ITU), 39
Internet service providers (ISPs), 103
Intrusion detection systems (IDSs), 73, 104
Investment factors, 222–223
Isolation Forest (IDS-IF), 104, 106
  abnormal data points, 107
  decisions trees, 107
  Euler's constant, 109
  experimental environment, 109–111
  Gini Index, 108
  intrusions/anomalies, 107
  multiple decision trees, 108
  performance evaluation, 111–115
  system architecture, 108

K
KDD dataset, 110–111
Key generation algorithm, 93
k-nearest neighbor (KNN), 105, 194, 208

L
Lagrange interpolation, 93
Lemmatization, 151
Linked data, 206–209, 213
Local Outlier Factor (LOF), 105, 106
Logistic Regression
  Arabic social media, hate speech detection in, 156–157
Long Short-Term Memory (LSTM), 61, 136
  Arabic social media, hate speech detection in, 158
  for real-time decision-making, 142–144
Long-Term Evolution (LTE)
  networks, 40, 87, 88
  standard, 88
Loss, 34, 35, 58, 60, 126
LSTM autoencoder, 221, 227

M
Machine learning (ML), 2, 119, 135
  algorithms, 192
  categories, 207–208
  drug discovery, QSAR anti-HIV feature selection and prediction, 194
  Electronic Health Records, 207, 208, 215
Machine learning and deep learning (ML/DL) techniques, 104
Machine learning for hate speech detection, 147–148
  classification performance scores, 159
  dataset, 160
    Alakrot's YouTube comments collection, 154–155
    YouTube, 153
  definition of, 148
  democratization of access, 149
  methodology
    evaluation metrics, 158–159
    Logistic Regression, 156–157
    Long Short-Term Memory, 158
    preprocessing, 155–156
    Random Forests, 157
    Support Vector Machines, 157

Machine learning for hate speech detection (*cont.*)
 MSA, 150
 natural language processing, 150–151
  Bag of Words and term frequency, 152
  stemming, 151
  term frequency-inverse document frequency, 152
 techniques, 149
Machine-Type Communications (MTC), 39–41
Manchester Triage Scale (MTS), 252
Manhattan distance, 253
 defining the problem, 254–255
 optimization framework, 255–256
Manual composition, 167
Market returns, 220
Markov Decision Process (MDP), 135
Medical data, 206–210
Medical ontologies, 206–208, 215
Medical RDF dataset, 206
Meta-model, 175–176
Meta-Object Facility (MOF) meta-model, 176
MIMO-FBMC/OQAM system, 22
 additive white Gaussian noise (AWGN) vector, 26
 as a BSS problem, 22
 discrete-time transmitted FBMC/OQAM signal, 23, 24
 DNN for blind detection and interference mitigation, 29–33
 FBMC/OQAM, 23, 25
 interference-free communication system, 22
 matricial model, 25
 simulation method
  bit error rate (35), 35–36
  parameters, 33
  RMSE and loss, 34–35
 SISO transmission systems, 24
 space–time block codes, 25
 using BSS formulation, 26–29
Minute Bar Trading Volumes, 229
Mobile ad hoc networks (MANET), 69
Modern Standard Arabic (MSA), 148, 150
Multi-access edge computing (MEC), 1, 6, 7, 9, 17
Multilinear regression (MLR), 200
Multiple access, 41
 See also Non-orthogonal multiple access (NOMA)
Multiple-input multiple-output (MIMO), 21
 antennas, 27
 with FBMC/OQAM, 21 (*see also* MIMO-FBMC/OQAM system)
 radio access techniques, 1
 as space–time block codes, 24

N
Naïve Bayes (NB), 157
Nash equilibrium, 80
Natural immune system (NIS), 250, 253
Natural Language Toolkit (NLTK), 155
Negative selection, 249
Network function virtualization (NFV), 1, 6, 7, 17
Network slicing, 2, 6
New Radio (NR) interface, 21, 40
Non-orthogonal multiple access (NOMA)
 DL-based NOMA, 50, 51
 for downlink, 42–45
 imperfection, 46–47
 methods, 40
 principle, 41–42
 spectral efficiency (SE) and energy efficiency (EE), 47–48
 system model, 41
 for uplink, 45–46
Non-repudiation, 90
NSL-KDD dataset, 106

O
Onboard units (OBUs), 85
Online BL-based Kalman filter (OBLKF), 22
OntoLearn ontology, 208
Ontology-based intention and context modeller, 176–177
Ontology Biomedical Repository, 211
OpenFlow (OF), 106
Orthogonal frequency-division multiple access (OFDMA), 41, 42, 44, 46, 48
Orthogonal frequency division multiplexing (OFDM), 22, 25, 41, 42, 44, 46, 48
Overstemming, 151
OWL interaction
 Planning Domain Description Language and
  mapping OWL, 182, 183
  planning and, 181–182
OWL-S extension
 for Semantic integration, 178–181

P
Part and Context Network (PCN), 137
Pedestrian Location Perception Network (P-LPN), 136, 145
 for pedestrian perception, 141–142

Peer-to-peer (P2P) oracle network, 13–14
Performance enhancements, 10–11
Performance evaluation, 99–101
Pervasive Information System (PIS), 165
Piotroski costumed factor, 240
"Plan Blanc", 247
Planning Domain Description Language (PDDL), 163
  CISCA architecture, 183
  context and context awareness, 164–165
  features, CISCA architecture, 183–184
  intention, 165–166
  intention and context modelling, 175
    context ontology, 177–178
    intention ontology, 177
    ontology-based intention and context modeller, 176–177
    proposed meta-model, 175–176
    semantic integration, OWL-S extension for, 178–181
  and OWL interaction
    mapping OWL, 182, 183
    planning and, 181–182
  remote accident management system, 187
  service composition, 166–167
    approaches, comparative table of, 172–174
    categories, 167
    technical classification of, 167–171
  service composition module, 185–186
Practical Byzantine Fault Tolerance (PBFT), 4
Private blockchain, 4
Process composition, 168
Proof of Activity (PoA), 4
Proof of Stake (PoS), 4
Proof of Work (PoW), 4
Protein sequence databases, 207
Public blockchain, 4

Q
Q-function, 131
Quantitative structure-activity relationships (QSAR) model, 192
Q-values, 130

R
RadLex, 208
Random Forests (RF)
  Arabic social media, hate speech detection in, 157
Ranked information coefficient (rank IC), 226
Reasoning based on rules (RBR), 246
Recurrent neural networks (RNNs), 49, 52, 53
  activation functions, 61
  advantages of, 59
  applications of, 60
  disadvantages of, 59–60
  loss function, 60
  LSTM, 61
  temporal back propagation, 61
  variants of, 58
Reinforcement learning (RL), 120–121, 130, 136
Reliable data transmission, 93
ReLU activation function, 227
Remote accident management system, 187
Resource Description Framework (RDF), 206–210, 213, 215
Return on assets (RoA), 240
Return on Attack (ROA), 70
Return on Investment (ROI), 70
Roadside units (RSU), 69
ROI and ROA, attack-defense tree, 74–76
Root mean square error (RMSE), 33–35

S
Scalable and secure blockchain for 5G
  architecture, 11
  Block5G, 11, 15
  data exchange in 5G network, 11, 12
  design components
    initialization, 15
    reward, 15–16
  P2P oracle network, 13–14
  shared blockchain, 12–13
Self-taught learning (STL-IDS), 106
Semantic approaches, 170
Semantic integration
  OWL-S extension for, 178–181
Semantic web, 207, 214
Service composition module, 185–186
ServiceGrounding, 179
ServiceModel, 179
ServiceProfile, 179
Setup algorithm, 93
Sharding, 5, 11–13, 19
Signal-to-interference ratio (SIR), 22
Signing algorithm, 93
SimpleProcess, 179
Singular value decomposition (SVD), 22
Smart contracts, 2, 5, 8, 10, 11, 17, 19
Social-distancing policies, 219, 220, 228
Software-defined network (SDN), 1, 5–6, 17, 104
Solidity, 5

Spanish Flu, 228
Spearman's correlation coefficient, 226
Spectral clustering algorithm, 221, 225
Static composition, 167
Stemming
  natural language processing, 151
Stock return, 220, 221, 233
Stress statuses, 247
Structural composition, 168–169
Successive Interference Cancellation (SIC), 41–43, 45–47, 50
Supervised machine learning models, 194
Supervision layer, CISCA architecture, 185
Support vector machines (SVM), 105, 106, 200–201
  Arabic social media, ML for hate speech detection in, 157
Susceptible–Infected–Recovered (SIR), 219

T
Task-Assistant CNN (TA-CNN), 136
Term frequency
  natural language processing, 152
Term frequency-inverse document frequency (TF-IDF), 152, 209, 210, 213, 215
Theme identification, 210
Third Generation Partnership Project (3GPP), 88
3G, 1, 39, 87
Traffic congestion, 119
Traffic sign detection
  accuracy curves and loss, 64
  autonomous cars, 53
  C-LSTM method, 54
  convolutional neural networks, 52, 53
    convolutional layers, 56
    correction layers (ReLU), 57–58
    feed-forward neural network, 55
    fully connected layers, 57
    learning phase, 55
    loss layer, 58
    multilayer neural networks, 55
    multiple filters, 57
    shared weights, 56
    structure of, 56
    subsampling layers, 57
    training process, 55
  DNN layer, 54
  LSTM layer, 54
  parameters configuration, 65
  proposed system, 61–62
  recurrent neural networks, 52, 53
    activation functions, 61
    advantages of, 59
    applications of, 60
    disadvantages of, 59–60
    loss function, 60
    LSTM, 61
    temporal back propagation, 61
    variants of, 58
  results obtained, 64–65
Twitter, 149

U
Ultra-reliable communications (URC), 1, 40
Ultra-reliable low-latency communication (URLLC), 21, 39
Understemming, 151
UniProt protein sequence databases, 207
User interface layer, CISCA architecture, 184
US firms in Healthcare and Utilities, 231

V
VANET availability
  attack-defense tree, 72–74
    black hole attack, 73
    denial of service attack, 73
    DOS attack, 73
    game theory, 76
    intrusion detection system, 73
    Nash equilibrium, 80
    players, 77
    reduction of, 79
    smurf attack, 73
    strategies, 77
    SYN cookies, 73
    transmitting dummy messages, 73
    utility matrix, 77, 78
  system model, 70
Vehicle to everything (V2X) 2 communication model
  access control, 91
  attribute-based signature (ABS), 84, 92–93
  authentication, 89
  availability, 90
  centralized architecture, 85, 86
  confidentiality, 90
  distributed architecture, 85–86
  elliptical curves, 91–92
  high-level protocol specification language (HLPSL), 95
  hybrid architecture, 86–87
  integrity, 89–90
  non-repudiation, 90
  performance evaluation, 99–101

reliable data transmission, 93
security analysis, 96–99
vehicular networks components
infrastructure, 85
onboard units (OBUs), 85
road users, 84–85
verification hypotheses, 95–96
V2I, 87–88
V2N, 87–88
V2P, 87–88
V2V, 87–88
Vehicle to Vehicle (V2V) communication, 85–86, 136
Vehicular ad hoc networks (VANET), 69
Verification algorithm, 93
Vulnerable Road Users (VRUs), 119–120, 135

W
W3C Resource Description Framework (RDF), 207
Web Ontology Language (OWL), 163, 207
Web Service Description Language (WSDL), 163
Web services, 163
White plan, 248, 260
WikiPathways, 207
Wireless communication systems, 21
World Health Organization (WHO), 135

Y
YouTube, 149, 153

CPSIA information can be obtained
at www.ICGtesting.com
Printed in the USA
LVHW081906010323
740706LV00010B/1174

9 783030 771874